Intelligence
is not Artificial

Why the Singularity is not coming any time soon and other Meditations on the Post-Human Condition and the Future of Intelligence

Piero Scaruffi

ISBN 978-0-9765531-9-9

To all the machines that will never become intelligent enough to understand the books that humans write about them and why we write them

Table of Contents

What this Book is about

Writers, inventors and entrepreneurs, impressed by progress in several scientific fields and notably in Artificial Intelligence, are debating whether we may be heading for a "singularity" in which machines with super-human intelligence will arise and multiply. At the same time, enthusiastic coverage in the media has widely publicized machines performing sophisticated tasks, from beating masters of go/weichi to driving a car, from recognizing cats in videos to outperforming human experts on TV quiz shows. These stories have re-ignited interest in Artificial Intelligence, whose goal is to create machines that are as intelligent as humans, and generated fears in the public that these intelligent machines might cause harm to humans, if not steal their jobs.

First of all, this book provides a "reality check" of sorts on Artificial Intelligence. I show that, in a society driven by news media that desperately need sensational news to make money and in an academic world increasingly driven by the desire to translate academic research into Silicon Valley start-ups, technological progress in general, and progress in computer science in particular, is often overrated. I wanted to dispel some notions and misconceptions, and my version of the facts may sound controversial until you read my explanations. I think that (real) progress in (real) Artificial Intelligence, since its founding, has been negligible, and one reason is, ironically, that computers have become so much more (computationally) powerful.

In general, we tend to exaggerate the uniqueness of our age, just as previous generations had done. The very premise of the singularity theory is that progress is accelerating like never before. I argue that there have been other eras of accelerating progress, and it is debatable if ours is truly so special. The less you know about the past the more you are amazed by the present.

There is certainly a lot of change in our era. But change is not necessarily progress, or, at least, it is not necessarily progress for everybody. Disruptive innovation is frequently more about disruption than about innovation because disruption creates huge new markets for the consumer electronics industry. This has nothing to do with machine intelligence, and sometimes not even much to do with innovation.

There is also an exaggerated metaphysical notion that human intelligence is some sort of an evolutionary climax. Maybe so, but it is worth cautioning that non-human intelligence is already among us, and is multiplying rapidly, but it is not a machine: countless animals are capable of feats that elude the smartest humans. For a long time we have also had machines capable of performing "superhuman" tasks. Think of the clock, invented almost 1,000 years ago, that can do something that no human can do: it can tell how many hours, minutes and seconds elapse between two events.

Once we realize that non-human intelligence has always been around, and that we were already building super-human machines centuries ago, the discussion about super-intelligent machines can be reframed in more historically and biologically meaningful terms.

The last generation or two missed out on the debates of the previous decades (the "Turing test", the "ghost in the machine", the "Chinese room", etc). Therefore it is much easier for new A.I. practitioners to impress the younger generations. I have summarized the various philosophical arguments in favor of and against the feasibility of machine intelligence in my book "Thinking about Thought" and i won't repeat them here. I will, however, at least caution the new generations that "grew up" (as far as cognitive science goes) at a time when the term "intelligence" was not "cool" at all: it was held to be too vague, too unscientific, too abused in popular literature to lend itself to scientific investigation. It is regrettable that it is being abused again, and, just like back then, without a proper definition of what we mean by "intelligence". Ask one hundred psychologists and you will get one hundred different definitions. Ask philosophers and you will get thick tomes written in a cryptic language. Ask neurobiologists and they may simply ignore you.

This is the mother of all problems in the debate on the "singularity": "singularity" and "superhuman intelligence" are non-scientific terms based on non-scientific coffee-house chatting.

The term "artificial intelligence" is even more confusing, a veritable moving target. In this book i capitalize Artificial Intelligence when i am referring to the discipline, while using lowercase "artificial intelligence" to refer to an intelligent machine or an intelligent software. A.I. practitioners also use the term "Artificial General Intelligence" (A.G.I.) to refer to a machine that will exhibit human-level intelligence, not just one intelligent skill.

I also feel that any discussion on machine intelligence should be complemented with an important (more important?) discussion about the changes in human intelligence due to the increased "intelligence" of machines. This change in human intelligence may have a stronger impact on the future of human civilization than the improvements in machine intelligence. To wit: the program of turning machines into humans is not very successful yet, but the program of turning humans into machines (via an endless repertory of rules and regulations) is very successful.

My perspective is a little different from the perspective of the many writers who have written, or are writing, books on Artificial Intelligence: i am a historian, not a futurist. I may not know the future, but at least i know the past.

I am intrigued by another sociological/anthropological aspect of this discussion: humans seem to have a genetic propensity to believe in higher forms of intelligence (gods, saints, UFOs, ...) and the Singularity (capitalized "S") could simply be its latest manifestation in our post-religious 21st century.

However, most people don't really care for how we call it: they are afraid not of some electromechanical monster that will kill the human race, but, quite simply, of losing their job to smarter and smarter machines. This too seems to me a wild exaggeration. New machines have always created new jobs, that are also better-paid. I fail to see why this time should be different. Remove all the sensational hyperboles, and it should be obvious that smarter machines will create more jobs, and better-paid jobs.

All of this explains why i am not afraid of Artificial Intelligence: 1. A reality check shows that most of its achievements are not that impressive; 2. Most of the "intelligence" displayed by machines is actually due to the structured environment that humans build for them; 3. The accelerating progress that we perceive is not unique in history; 4. We have always been surrounded by super-human (or, better, non-human) intelligence; 5. I am more concerned about the future of human intelligence than about the future of machine intelligence.

We actually need intelligent machines. Technological progress has solved many problems, but there are still people dying of diseases and dangerous jobs, and we will soon have an ageing society that will need even more help from technology. I am not afraid that "intelligent" machines are coming. I am afraid that they will not come soon enough.

This book was started in September 2013 and this revised edition was completed in June 2016.

piero scaruffi

P.S.: Yes, i don't like to capitalize the first person pronoun "i".

Sociological Background

Historians, scientists, philosophers and poets alike have written that the human being strives for the infinite. In the old days this meant (more or less) that s/he strives to become one with the god who created and rules the world. As atheism began to make strides in Western civilization, Arthur Schopenhauer rephrased the concept as a "will to power". Friedrich Nietzsche confirmed that the god of the Bible is dead, and the Western search for the "infinite" became a mathematical and scientific program instead of a mystical one. About a century ago, European mathematicians such as Bertrand Russell and David Hilbert launched a logical program that basically aimed at making it easy to prove and discover everything that can be. The perspective therefore changed: instead of something that humans have to attain, the infinite has become something that humans will build.

One of the consequences of this line of research was the creation of the digital electronic computer, the physical implementation of a thought experiment by the British mathematician Alan Turing. He also wrote a pioneering paper about machine intelligence ("Computing Machinery and Intelligence", 1950) and a few years later the term "artificial intelligence" was already popular among both scientists and philosophers. The first conference on Artificial Intelligence was held in 1956 at Dartmouth College in New Hampshire, organized by John McCarthy with help from MIT scientist Marvin Minsky and others. That was just ten years after the introduction of the first general-purpose computer, the ENIAC. From the very beginning, the electronic computer had been dubbed by the media "the electronic brain".

The idea behind the Singularity, a concept popularized by Ray Kurzweil's "The Age of Intelligent Machines" (1990) and by his subsequent, highly-successful, public-relations campaign, is that we are about to witness the advent of machines that are more intelligent than humans, so intelligent that humans can neither control them nor understand them.

Admittedly, the discipline of Artificial Intelligence, that had largely languished in the 1990s and 2000s, has staged a revival of sorts, both in the eyes of the public and in the eyes of big business. Achievements in the field of A.I. are often hailed by the mainstream media as steps towards machine domination, and investment in A.I. startups has multiplied several times over to reach record levels.

In the age that has seen the end of human space exploration, the retirement of the only commercial supersonic airplane, the decline of nuclear power, and the commercialization of the Internet (an event that basically turned a powerful scientific tool into a marketing tool and a form of light entertainment), machine intelligence seems to bring some kind of collective reassurance that we are not, after all, entering a new Dark Age; on the contrary, we are witnessing the dawn of a superhuman era. Of course, decades of science fiction books and movies helped create the ideal audience for this kind of scenario.

However, the tone and the (very weak) arguments in favor of the Singularity do remind one of religious prophecies, except that this time the coming messiah will be a product made by us instead of being sent by an external divinity. In a sense, this is a religion according to which we are creating the divinity.

The idea of the Singularity is fascinating because it plays the history of religion backwards. Religion traditionally is meant to explain the mystery of the complexity of the universe, the miracle of life, the purpose of consciousness. Even some of today's eminent scientists subscribe to the "creationist" notion that a superhuman intelligence was required to create the world. This theory is frequently called "intelligent design" but it would be more appropriate to call it "super-intelligent design" because "intelligent" only refers to human intelligence. The whole point of religion was precisely to posit the existence of something that human intelligence could never possibly build. The hidden assumption of religion is that all the laws of nature that humans can possibly discover will never be enough to explain the mysteries of the universe, of life, of the soul. Whatever can be explained by those mathematical laws can also be implemented by humans, and therefore does not require the existence of supernatural forces. God, instead, is a singularity, the singularity that preceded human intelligence and created it, and is infinitely superior to it.

Luckily for us, this supreme deity is also capable of, and somewhat interested in, granting us immortality, which, at the end of the day, is what believers hope to obtain from believing.

Today's hypothesis of a coming singularity due to super-intelligent machines provides a mirror image of this story. The original singularity (God) was needed to explain the inexplicable. The new singularity (the machine) will be unexplainable. Human intelligence could not, in the past, understand the nature of a God who created human intelligence; nor can it, in the present, understand the workings of the super-intelligent machine of the future that human intelligence will have created.

The Singularity movement is split into two camps: the optimists and the pessimists. The optimists think that machines will make us immortal. The pessimists think that machines will kill us all. I still haven't heard anyone take the kind of intermediary position that most religions take: good people will go to paradise, bad people will go to hell. Apparently, the Singularity will not distinguish between good and bad people: either it kills everybody or it makes everybody immortal. (Money may be more important than good deeds for one to become immortal because, if I understand correctly, immortality will be a service available for sale or for rent just like cloud computing is today).

"This is the whole point of technology. It creates an appetite for immortality on the one hand. It threatens universal extinction on the other. Technology is lust removed from nature.... It's what we invented to conceal the terrible secret of our decaying bodies". (Don DeLillo, "White Noise")

It is sometimes difficult to argue with the Singularity crowd because they often seem unaware that some of the topics they discuss have been discussed for a long time, with pros and cons, by philosophers and scientists. In its worst manifestation the Singularity movement is becoming the religion of high-tech nerds who did not study history, philosophy, or science, not even computer science. At its best, however, it helps acquaint the general public with a society of (software and hardware) robots that is inevitably coming, although its imminence might be wildly exaggerated.

It may not be a coincidence that the boom of interest in the Singularity originated in the USA (a country well acquainted with apocalyptic evangelism, conspiracy theories, UFO sightings and cryptic prophets like Nostradamus) and that it originated after the year 2,000, a year that had three zeroes according to the calendar introduced by Pope Gregory in 1582 and that, because of those three zeroes, was thought by many to herald a major discontinuity in history if not the end of the world itself. For a while the world was shaken almost yearly by catastrophic predictions, most famously (in the USA) Harold Camping's Biblical calculations that the end of the world was coming on the 21st of October of 2011 and the theory that the end of the Mayan calendar (December 21, 2012) marked the end of the world. Luckily, they were all proven wrong, but maybe they created a public opinion ready to be fascinated by a technological version of the same general story (the end of the human race).

Irony aside, it is fascinating to see how religion is being reinvented on completely different foundations in Silicon Valley.

A Brief History of Artificial Intelligence/ Part 1

One can start way back in the past with the ancient Greek and Chinese automata of two thousand years ago, or with the first electromechanical machines of a century ago, but to me a history of machine intelligence begins in earnest with the "universal machine", originally conceived in 1936 by the British mathematician Alan Turing. He did not personally build it, but Turing realized that one could create the perfect mathematician by simulating the way logical problems are solved: by manipulating symbols. The first computers were not Universal Turing Machines (UTM), but most computers built since the ENIAC (1946), including all the laptops and smartphones that are available today, are UTMs. Because it was founded on predicate logic, which only admits two values ("true" and "false"), the computer at the heart of any "intelligent" machine relies on binary logic (ones and zeroes).

Cybernetics (that can be dated back to the 1943 paper "Behavior, Purpose and Teleology" co-written by MIT mathematician Norbert Wiener, physiologist Arturo Rosenblueth and engineer Julian Bigelow) did much to show the relationship between machines and living organisms. One can argue that machines are a form of life or, vice versa, that living organisms are forms of machinery.

However, "intelligence" is commonly considered one or many steps above the merely "alive": humans are generally considered intelligent (by fellow humans), whereas worms are not.

The "Turing Test", introduced by the same Alan Turing in his paper "Computing Machinery and Intelligence" (1950), has often been presented as the kind of validation that a machine has to pass in order to be considered "intelligent": if a human observer, asking all sorts of questions, cannot tell whether the agent providing the answers is human or mechanical, then the machine has become intelligent (or, better, as intelligent as the human being). The practitioners of Artificial Intelligence quickly split into two fields. One, pioneered by Herbert Simon and his student Allen Newell at Carnegie Mellon University with their "Logic Theorist" (1956), basically understood intelligence as the pinnacle of mathematical logic, and focused on symbolic processing. In 1959 Arthur Samuel at IBM in New York wrote not only the first computer program that could play checkers but the first self-learning program. That program implemented the alpha-beta search algorithm, that would dominate A.I. for the next 20 years. The first breakthrough in this branch of A.I. was probably John McCarthy's article "Programs with Common Sense" (1959): McCarthy understood that someday machines would easily be better than humans at many repetitive and computational tasks, but "common sense" is what really makes someone "intelligent" and common sense comes from knowledge of the world. That article spawned the discipline of "knowledge representation": how can a machine learn about the world and use that knowledge to make inferences. This approach was somehow "justified" by the idea, introduced by the MIT linguist Noam Chomsky in "Syntactic Structures" (1957), that language competence is due to some grammatical rules that express which sentences are correct in a language. The grammatical rules express "knowledge" of how a language works, and, once you have that knowledge (and a vocabulary), you can produce any sentence in that language, including sentences you have never heard or read before.

The rapid development of computer programming helped this field take off, as computers were getting better and better at processing symbols: knowledge was represented in symbolic structures and "reasoning" was reduced to a matter of processing symbolic expressions. This line of research led to "knowledge-based systems" (or "expert systems"), such as Ed Feigenbaum's Dendral (1965) at Stanford, that consisted of an "inference engine" (the repertory of legitimate reasoning techniques recognized by the mathematicians of the world) and a "knowledge base" (the "common sense" knowledge). This technology relied on acquiring knowledge from domain experts in order to create "clones" of such experts (machines that performed as well as the human experts). The limitation of expert systems was that they were "intelligent" only in one specific domain.

Meanwhile, the other branch of Artificial Intelligence was pursuing a rather different approach: simulating what the brain does at the physical level of neurons and synapses. The logical school of John McCarthy and Marvin Minsky believed in using mathematical logic to simulate how the human mind works; the school of "neural networks" (or "connectionism") believed in simulating the structure of the brain in order to simulate how the brain works.

Since in the 1950s neuroscience was just in its infancy (medical machines to study living brains would not become available until the 1970s), computer scientists only knew that the brain consists of a huge number of interconnected neurons, and neuroscientists were becoming ever more convinced that "intelligence" was due to the connections, not to the individual neurons. A brain can be viewed as a network of interconnected nodes, and our mental life as due to the way messages travel through those connections from the neurons of the sensory system up to the neurons that process those sensory data and eventually down to the neurons that generate action. The neural connections can vary in strength from zero to infinite. Change the strength of some neural connections and you change the outcome. In other words, the strength of the connections can be tweaked to cause different outputs for the same inputs. The problem for those designing "neural networks" consists in fine-tuning the connections so that the network as a whole comes up with the correct interpretation of the input; e.g. with the word "apple" when the image of an apple is presented. This is called "training the network". For example, showing many apples to the system and forcing the answer "APPLE" should result in the network adjusting those connections to recognize apples. This is called "supervised learning". Since the key is to adjust the strength of the connections, the alternative term for this branch of A.I. is "connectionism". Frank Rosenblatt's Perceptron (1957) at Cornell University and Oliver Selfridge's Pandemonium (1958) at the MIT were the pioneer "neural networks": not knowledge representation and logical inference, but pattern propagation and automatic learning. Compared with expert systems, neural networks are dynamic systems (their configuration changes as they are used) and predisposed to learning by themselves (they can adjust their configuration). "Unsupervised" networks, in particular, can discover categories by themselves; e.g., they can discover that several images refer to the same kind of object, a cat.

There are two ways to solve a crime. One way is to hire the smartest detective in the world, who will use experience and logic to find out who did it. On the other hand, if we had enough surveillance cameras placed around the area, we would scan their tapes and look for suspicious actions. Both ways may lead to the same conclusion, but one uses a logic-driven approach (symbolic processing) and the other one uses a data-driven approach (ultimately, the visual system, which is a connectionist system).

In 1969 Marvin Minsky and Samuel Papert of the MIT published a devastating critique of neural networks (titled "Perceptrons") that virtually killed the discipline. At the same time expert systems were beginning to make inroads at least in academia, notably Bruce Buchanan's Mycin (1972) at Stanford for medical diagnosis and John McDermott's Xcon (1980) at Carnegie Mellon University for product configuration, and, by the 1980s, also in the industrial and financial worlds at large, thanks especially to many innovations in knowledge representation (Ross Quillian's semantic networks at Carnegie Mellon University, Minsky's frames at the MIT, Roger Schank's scripts at Yale University, Barbara Hayes-Roth's blackboards at Stanford University, etc). Intellicorp, the first major start-up for Artificial Intelligence, was founded in Silicon Valley in 1980. There was progress in knowledge-based architectures to overcome the slow speed of computers. In 1980 Judea Pearl introduced the Scout algorithm, the first algorithm to outperform alpha-beta, and in 1983 Alexander Reinefeld further improved the search algorithm with his NegaScout algorithm.

One factor that certainly helped the symbolic-processing approach and condemned the connectionist approach was that the latter uses much more complex algorithms, i.e. it requires computational power that at the time was rare and expensive.

(Personal biography: i entered the field in 1985 and went on to lead the Silicon Valley-based Artificial Intelligence Center of the largest European computer manufacturer, Olivetti, and i later worked at Intellicorp for a few years).

Footnotes in the History of Artificial Intelligence

There were many side tracks that didn't become as popular as expert systems and neural networks.

At the famous conference on A.I. of 1956 there was a third proposal for A.I. research. The Boston-based mathematician Ray Solomonoff presented "An Inductive Inference Machine" for machine learning. Induction is the kind of learning that allows us to apply what we learned in one case to other cases. His method used Bayesian reasoning, i.e. it introduced probabilities in machine learning. Alas, Solomonoff's inductive inference is not computable, although some algorithms can approximate it in order to make it run on a computer.

The robot Shakey (1969), built at the Stanford Research Institute (SRI) by Charles Rosen's team, was the vanguard of autonomous vehicles.

Cordell Green experimented at Stanford with automatic programming, software that can write software the same way a software engineer does ("Application of Theorem Proving to Problem Solving", 1969).

In 1961 Melvin Maron, a philosopher working at the RAND Corporation, suggested a statistical approach to analyze language (technically speaking, a "naive Bayes classifier"). IBM's Shoebox (1964) debuted speech recognition. Conversational agents such as Joe Weizenbaum's Eliza (1966) and Terry Winograd's Shrdlu (1972), both from the MIT, were the first practical implementations of natural language processing and conversational agents.

In 1968 Peter Toma founded Systran to commercialize machine-translation systems. The discipline of Machine Translation actually predates Artificial Intelligence. Yehoshua Bar-Hillel organized the first International Conference on Machine Translation in 1952 at the MIT. In 1954 Leon Dostert's team at Georgetown University and Cuthbert Hurd's team at IBM demonstrated a machine-translation system, one of the first non-numerical applications of the digital computer. (For the record, in 1958 the same Bar-Hillel who had jumpstarted the field published a "proof" that machine translation is impossible without common-sense knowledge).

Refining an idea pioneered by the German engineer Ingo Rechenberg at the Technical University of Berlin in his thesis "Evolution Strategies" (1971), John Holland at the University of Michigan introduced a different way to construct programs by using "genetic algorithms" (1975), the software equivalent of the rules used by biological evolution: instead of writing a program to solve a problem, let a population of programs evolve (according to some algorithms) to become more and more "fit" (better and better at finding solutions to that problem). In 1976 Richard Laing at the same university introduced the paradigm of self-replication by self-inspection ("Automaton Models of Reproduction by Self-inspection") that 27 years later would be employed by Jackrit Suthakorn and Gregory Chirikjian at Johns Hopkins University to build a rudimentary self-replicating robot ("An Autonomous Self-Replicating Robotic System", 2003).

In 1990 Carver Mead at Caltech described a "neuromorphic" processor, a processor that emulates the human brain.

A Premise to the History of Artificial Intelligence

Surprisingly few people ask "why?" Why did the whole program of A.I. get started in the first place? What is the goal? Why try and build a machine that behaves (and feels?) like a human being?

There were and there are several motivations. I believe the very first spark was pure scientific curiosity. A century ago an influential German mathematician, David Hilbert, outlined a program to axiomatize mathematics as a sort of challenge for the world's mathematicians. In a sense, he asked if we can discover a procedure that will allow anybody to solve any mathematical problem: run that procedure and it will prove any theorem. In 1931 Kurt Goedel proved his Theorem of Incompleteness, which was a response to Hilbert's challenge. It concluded: "No, that's not possible, because there will always be at least one proposition that we cannot prove true or false"; but in 1936 Alan Turing offered his solution, now known as the Universal Turing Machine, which is as close as we can get to Hilbert's dream procedure. Today's computers, including your laptop, your notepad and your smartphone, are Universal Turing Machines. And then the next step was to wonder if that machine can be said to be "intelligent", i.e. can behave like a human being (Turing's Test), can have conscious states, and can be even smarter than its creator (the Singularity).

The second motivation was purely business. Automation has been a source of productivity increase and wealth creation since ancient times. The rate of automation accelerated during the industrial revolution and it still is an important factor in economic development. There isn't a day when a human being isn't replaced by a machine. Machines work 24 hours a day and 7 days a week, don't go on strike, don't have to stop for lunch, don't have to sleep, don't get sick, don't get angry or sad. Either they function or they don't. If they don't, we simply replace them with other machines. Automation was pervasive in the textile industry way before computers were invented. Domestic appliances like dishwashers automated household chores. Assembly lines automated manufacturing. Agricultural machines automated grueling rural chores. That trend continues. As i type, machines (sensing cameras hanging from traffic lights remotely connected to the traffic division of a city) are replacing traffic police in many cities of the world to direct traffic (and to catch drivers who don't stop at red lights).

A third motivation was idealistic. An "expert system" could provide the service that the best expert in the world provides. The difference is that the human expert cannot be replicated all over the world, the expert system could. Imagine if we had an expert system that clones the greatest doctors in the world such that we could make that expert system available for free to the world's population (rich or poor), 24 hours a day, 7 days a week.

Brain Simulation and Intelligence

Behind the approach of neural networks is the hidden assumption that intelligence, and perhaps consciousness itself, arises out of complexity. This is a notion that dates back at least to the British neurophysiologist William Grey-Walter who in 1949, before the age of digital computers, was already designing early robots named Machina Speculatrix using analogue electronic circuits to simulate brain processes. More recently, David Deamer at UC Santa Cruz has calculated the "brain complexity" of several animals ("Consciousness and intelligence in mammals: Complexity thresholds", 2012).

All the "intelligent" brains that we know are made of neurons. Could the brain be made of ping-pong balls and still be as intelligent? If we connect a trillion ping-pong balls do we get a conscious being? What if the ping-pong balls are made of a material that conducts electricity? What if i connect them exactly like the neurons are connected in my brain: do i get a duplicate of my consciousness or at least a being that is as "intelligent" as me? The hidden assumption behind neural networks is that the material doesn't matter, that it doesn't have to be neurons (flesh), at least insofar as intelligence is concerned; hence, a purist of connectionism would argue that a system made of a trillion ping-pong balls would be as intelligent as me, as long as it duplicates exactly what happens in my brain.

The Body

A lot of what books on machine intelligence say is based on a brain-centered view of the human being. I may agree that my brain is the most important organ of my body (i'm ok with transplanting just about any organ of my body except my brain). However, this is probably not what evolution had in mind. The brain is one of the many organs designed to keep the body alive so that the body can find a mate and make children. The brain is not the goal but one of the tools to achieve that goal.

Focusing only on mental activities when comparing humans and machines is a categorical mistake. Humans do have a brain but don't belong to the category of brains: they belong to the category of animals, which are mainly recognizable by their bodies. Therefore, one should compare machines and humans based on bodily actions and not just on the basis of printouts, screenshots and files. Playing a match of chess with the world champion of chess is actually easy. It is much harder for a machine to do any of the things that we routinely do in our home (that our bodies do). Playing chess is actually much easier than playing soccer with a group of children.

Furthermore, there's the meaning of action. The children who play soccer actually enjoy it. They scream, they are competitive, they cry if they lose, they can be mean, they can be violent. There is passion in what we do. Will an android that plays decent soccer in 3450 (that's a realistic date in my opinion) also have all of that? Let's take something simpler, that might happen in 50 or 100 years: at some point we'll have machines capable of reading a novel; but will they understand what they are reading? Is it the same "reading" that i do? This is not only a question about the self-awareness of the machine but about what the machine will do with the text it reads. I can find analogies with other texts, be inspired to write something myself, send the text to a friend, file it in a category that interests me. There is a follow-up to it. Machines that read a text and simply produce an abstract representation of its content (and we are very far from the day when they will be able to do so) are useful only for the human who will use it.

The same considerations apply to all the corporeal activities that are more than simple movements of limbs.

The body is the reason why i think the Turing Test is not very meaningful. The Turing Test locks a computer and a human being in two rooms, and, by doing so, it removes the body from the test. My test (let's immodestly call it the Scaruffi Test) would be different: we give a soccer ball to both the robot and the human and see who dribbles better. I am not terribly impressed that a computer beat the world champion of chess (i am more impressed with the human, that it took so long for a machine with virtually infinite memory and processing power to beat a human). I will be more impressed the day a robot dribbles better than Lionel Messi.

If you remove the body from the Turing test, you are removing pretty much everything that defines a human being as a human being. A brain kept in a jar is not a human being: it is a gruesome tool for classrooms of anatomy.

(I imagine my friends at the nearest A.I. lab already drawing sketches of robots capable of intercepting balls and then kicking them with absolute precision towards the goal and with such force that no goalkeeper could catch them; but that's precisely what we don't normally call "intelligence", that is precisely what clocks and photocopiers do, i.e. they can do things that humans cannot do such as keeping accurate time and making precise copies of documents, and that is not yet what Messi does when he dribbles defenders).

Intermezzo: We may Overestimate Brains

The record for brain size compared with body mass does not belong to Homo Sapiens: it belongs to the squirrel monkey (5% of the body weight, versus 2% for humans). The sparrow is a close second.

The longest living beings on the planet (bacteria and trees) have no brain.

Childhood

A machine is born adult. It's the equivalent of you being born at 25; and never aging until the day that an organ stops working. One of the fundamental facts of human intelligence is that it comes via childhood. First we are children, then we get the person that is writing (or reading) these lines. The physical development of the body goes hand in hand with the cognitive development of the mind. The developmental psychologist Alison Gopnik has written in "The Philosophical Baby" (2009) that the child's brain is wildly different from the adult brain (in particular the prefrontal cortex). She even says that they are two different types of Homo Sapiens, the child and the adult. They physically perform different functions. Whether it is possible to create "intelligence" equivalent to human intelligence without a formation period is a big unknown.

Alison Gopnik emphasized the way children learn about both the physical world and the social world via a process of "counterfactuals" (what ifs): they understand how the (physical and psychological) worlds function, then they create hypothetical ones (imaginary worlds and imaginary friends), then they are ready to create real ones (act in the world to change it and act on people to change their minds). When we are children, we learn to act "intelligently" on both the world and on other people. Just like everything else with us, this function is not perfect. For example, one thing we learn to do is to lie: we lie in order to change the minds around us. A colleague once proudly told me: "Machines don't lie." That is one reason why i think that science is still so far away from creating intelligent machines. To lie is something you learn to do as a child, among many other things, among all the things that are our definition of "intelligence".

Deep Learning - A brief History of Artificial Intelligence/ Part 2

Knowledge-based systems did not expand as expected: the human experts were not terribly excited at the idea of helping construct clones of themselves, and, in any case, the clones were not terribly reliable.

Expert systems also failed because of the World-wide Web: you don't need an expert system when thousands of human experts post the answer to all possible questions. All you need is a good search engine. That search engine plus those millions of items of information posted (free of charge) by thousands of people around the world do the job that the "expert system" was supposed to do.

The expert system was a highly intellectual exercise in representing knowledge and in reasoning heuristically. The Web is a much bigger knowledge base than any expert-system designer ever dreamed of. The search engine has no pretense of sophisticated logic but, thanks to the speed of today's computers and networks, it "will" find the answer on the Web. Within the world of computer programs, the search engine is a brute that can do the job once reserved to artists.

Note that the apparent "intelligence" of the Web (its ability to provide all sorts of questions) arises from the "non-intelligent" contributions of thousands of people in a way very similar to how the intelligence of an ant colony emerges from the non-intelligent contributions of thousands of ants.

In retrospect a lot of sophisticated logic-based software had to do with slow and expensive machines. As machines get cheaper and faster and smaller, we don't need sophisticated logic anymore: we can just use fairly dumb techniques to achieve the same goals. As an analogy, imagine if cars, drivers and gasoline were very cheap and goods were provided for free by millions of people: it would be pointless to try and figure out the best way to deliver a good to a destination because one could simply ship many of those goods via many drivers with an excellent chance that at least one good would be delivered on time at the right address. The route planning and the skilled knowledgeable driver would become useless, which is precisely what has happened in many fields of expertise in the consumer society: when is the last time you used a cobbler or a watch repairman?

The motivation to come up with creative ideas for A.I. scientists was due to slow, big and expensive machines. Now that machines are fast, small and cheap, the motivation to come up with creative ideas is much reduced. Now the real motivation for A.I. scientists is to have access to thousands of parallel processors and let them run for months. Creativity has shifted to coordinating those processors so that they will search through billions of items of information. The machine intelligence required in the world of cheap computers has become less of a logical intelligence and more of a "logistical" intelligence.

Meanwhile, in the 1980s some conceptual breakthroughs fueled real progress in robotics. The Italian cyberneticist Valentino Braitenberg, in his "Vehicles" (1984), showed that no intelligence is required for producing "intelligent" behavior: all that is needed is a set of sensors and actuators. As the complexity of the "vehicle" increases, the vehicle seems to display an increasingly intelligent behavior. Starting in about 1987, Rodney Brooks at the MIT began to design robots that use little or no representation of the world. One can know nothing, and have absolutely no common sense, but still be able to do interesting things if equipped with the appropriate set of sensors and actuators.

The 1980s also witnessed a progressive rehabilitation of neural networks, a process that turned exponential in the 2000s. The discipline was rescued in 1982 by the CalTech physicist John Hopfield, who described a new generation of neural networks, based on simulating the physical process of annealing. These neural networks were immune to Minsky's critique. Hopfield's key intuition was to note the similarity with statistical mechanics. Statistical mechanics translates the laws of Thermodynamics into statistical properties of large sets of particles. The fundamental tool of statistical mechanics (and soon of this new generation of neural networks) is the Boltzmann distribution (actually discovered by Josiah-Willard Gibbs in 1901), a method to calculate the probability that a physical system is in a specified state. Meanwhile, in 1974, Paul Werbos had worked out a more efficient way to train a neural network: the "backpropagation" algorithm.

Building on Hopfield's ideas, in 1983 Geoffrey Hinton at Carnegie Mellon University and Terry Sejnowski at John Hopkins University developed the so-called Boltzmann machine (technically, a Monte Carlo version of the Hopfield network), a software technique for networks capable of learning; and in 1986 Paul Smolensky at the University of Colorado introduced a further optimization, the Restricted Boltzmann Machine. These were carefully calibrated mathematical algorithms to build neural networks to be both feasible (given the dramatic processing requirements of neural network computation) and plausible (that solved the problem correctly). Historical trivia: the Monte Carlo method of simulation had been one of the first applications that John Von Neumann had programmed in the ENIAC, right after inventing it with Stanislaw Ulam in 1946 as part of a top-secret military program.

This school of thought merged with another one that was coming from a background of statistics and neuroscience. Credit goes to Judea Pearl of UC Los Angeles for introducing Bayesian thinking into Artificial Intelligence to deal with probabilistic knowledge ("Reverend Bayes on Inference Engines", 1982). Thomas Bayes was the 18th century mathematician who developed Probability Theory as we know it today. Ironically, he never published his main achievement, that today we know as Bayes' theorem.

A kind of Bayesian network, the Hidden Markov Model, was already being used by A.I., particularly for speech recognition. The Hidden Markov Model is a Bayesian network that has the sense of time and can model a sequence of events. It was invented by Leonard Baum in 1966 at the Institute for Defense Analyses in New Jersey.

Reinforcement learning was invented even before the field was called "Artificial Intelligence": it was the topic of Minsky's PhD thesis in 1954. Reinforcement learning was first used in 1959 by Samuel's checkers-playing program. In 1961 British wartime code-breaker, Alan Turing cohort and molecular biologist Donald Michie at the University of Edinburgh built a device (made of matchboxes!) to play Tic-Tac-Toe called MENACE (Matchbox Educable Noughts and Crosses Engine) that learned how to improve its performance. In 1976 John Holland (of genetic algorithms fame) introduced classifier systems, which are reinforcement-learning systems. All the studies on reinforcement learning since Michie's MENACE converged together in the Q-learning algorithm invented in 1989 at Cambridge University by Christopher Watkins, which was, technically speaking, a Markov decision process ("Learning from

Delayed Rewards", 1989). Watkins basically discovered the similarities between reinforcement learning and the theory of optimal control that had been popular in the 1950s thanks to the work of Lev Pontryagin in Russia (the "maximum principle" of 1956) and Richard Bellman at RAND Corporation (the "Bellman equation" of 1957). Trivia: Bellman is the one who coined the expression "the curse of dimensionality" that came to haunt the field of neural networks.

Meanwhile, the Swedish statistician Ulf Grenander (who in 1972 had established the Brown University Pattern Theory Group) fostered a conceptual revolution in the way a computer should describe knowledge of the world: not as concepts but as patterns. His "general pattern theory" provided mathematical tools for identifying the hidden variables of a data set. Grenander's pupil David Mumford studied the visual cortex and came up with a hierarchy of modules in which inference is Bayesian, and it is propagated both up and down ("On The Computational Architecture Of The Neocortex II", 1992). The assumption was that feedforward/feedback loops in the visual region integrate top-down expectations and bottom-up observations via probabilistic inference. Basically, Mumford applied hierarchical Bayesian inference to model how the brain works.

Hinton's Helmholtz machine of 1995 was de facto an implementation of those ideas: an unsupervised learning algorithm to discover the hidden structure of a set of data based on Mumford's and Grenander's ideas.

The hierarchical Bayesian framework was later refined by Tai-Sing Lee of Carnegie Mellon University ("Hierarchical Bayesian Inference In The Visual Cortex", 2003). These studies were also the basis for the widely-publicized "Hierarchical Temporal Memory" model of the startup Numenta, founded in 2005 in Silicon Valley by Jeff Hawkins, Dileep George and Donna Dubinsky; yet another path to get to the same paradigm: hierarchical Bayesian belief networks.

The field did not take off until 2006, when Geoffrey Hinton at the the University of Toronto developed Deep Belief Networks, a fast learning algorithm for Restricted Boltzmann Machines. What had truly changed between the 1980s and the 2000s was the speed (and the price) of computers. Hinton's algorithms worked wonders when used on thousands of parallel processors. That's when the media started publicizing all sorts of machine-learning feats.

Deep Belief Networks are layered hierarchical architectures that stack Restricted Boltzmann Machines one on top of the other, each one feeding its output as input to the one immediately higher, with the two top layers forming an associative memory. The features discovered by one RBM become the training data for the next one.

Hinton and others had discovered how to create neural networks with many layers. One layer learns something and passes it on to the next one, which uses that something to learn something else and passes it on to the next layer, etc.

DBNs are still limited in one respect: they are "static classifiers", i.e. they operate at a fixed dimensionality. However, speech or images don't come in a fixed dimensionality, but in a (wildly) variable one. They require "sequence recognition", i.e. dynamic classifiers, that DBNs cannot provide. One method to expand DBNs to sequential patterns is to combine deep learning with a "shallow learning architecture" like the Hidden Markov Model.

Another thread in "deep learning" originated with convolutional networks invented in 1980 by Kunihiko Fukushima in Japan. Fukushima's Neocognitron was directly based on the studies of the cat's visual system published in 1962 by two Harvard neurobiologists, David Hubel (originally from Canada) and Torsten Wiesel (originally from Sweden). They proved that visual perception is the result of successive transformations, or, if you prefer, of propagating activation patterns. They discovered two types of neurons: simple cells, which respond to only one type of visual stimulus and behave like convolutions, and complex cells. Fukushima's system was a multi-stage architecture that mimicked those different kinds of neurons.

In 1989 Yann LeCun at Bell Labs applied backpropagation to convolutional networks to solve the problem of recognizing handwritten numbers (and then in 1994 for face detection and then in 1998 for reading cheques).

Deep neural networks already represented progress over the traditional three-layer networks, but it was really the convolutional approach that made the difference. They are called "convolutional" because they employ a technique of filtering that recalls the transformations caused by the mathematical operation of convolution.

A convolutional neural network consists of several convolutional layers. Each convolution layer consists of a convolution or filtering stage (the "simple cell"), a detection stage, and a pooling stage (the "complex cell"), and the result of each convolutional layer is in the form of "feature maps", and that is the input to the next convolutional layer. The last layer is a classification module.

The detection stage of each convolutional layer is the middleman between simple cells and complex cells and provides the nonlinearity of the traditional multi-layer neural network. Traditionally, this nonlinearity was provided by a mathematical function called "sigmoidal", but in 2011 Yoshua Bengio ("Deep Sparse Rectifier Networks") introduced a more efficient function, the "rectified linear unit", also inspired by the brain, that have the further advantage of avoiding the "gradient vanishing" problem of sigmoidal units.

Every layer of a convolutional network detects a set of features, starting with large features and moving on to smaller and smaller features. Imagine a group of friends subjected by you to a simple game. You show a picture to one of them, and allow him to provide a short description of the picture to only another one and using only a very vague vocabulary; for example: an object with four limbs and two colors. This new person can then summarize that description in a more precise vocabulary to the next person; for example a four-legged animal with black and white stripes. Each person is allowed to use a more and more specific vocabulary to the next person. Eventually, the last person can only utter names of objects, and hopefully correctly identifies the picture because, by the time it reaches this last person, the description has become fairly clear (e.g. the mammal whose skin is black and white, i.e. the zebra).

(Convolution is a well-defined mathematical operation that, given two functions, generates a third one, according to a simple formula. This is useful when the new function is an approximation of the first one, but easier to analyze. You can find many websites that provide "simple" explanations of what a convolution is and why we need them: these "simple" explanations are a few pages long, and virtually nobody understands them, and each of them is completely different from the other one. Now you know where the term "convoluted" comes from!)

In 1990 Robert Jacobs at the University of Massachusetts introduced the "mixture-of-experts" architecture that trains different neural networks simultaneously and lets them compete in the task of learning, with the result that different networks end up learning different functions ("Task Decomposition Through Competition in a Modular Connectionist Architecture", 1990).

Meanwhile in 1996 David Field and Bruno Olshausen at Cornell University had invented "sparse coding", an unsupervised technique for neural networks to learn the patterns inherent in a dataset. Sparse coding helps neural networks represent data in an efficient way that can be used by other neural networks.

Each neural network is, ultimately, a combination of "encoder" and "decoder": the first layers encode the input and the last layers decode it. For example, when my brain recognizes an object as an apple, it has first encoded the image into some kind of neural activity (representing shape, color, size, etc of the object), and has then decoded that neural activity as an apple.

The "stacked auto-encoders" developed in 2007 by Yoshua Bengio at the University of Montreal further improved the efficiency of capturing patterns in a dataset. There are cases in which a neural network would turn into a very poor classifier because of the nature of the training data. In that case a neural network called "autoencoder" can learn the important features in an unsupervised way. So autoencoders are special cases of unsupervised neural networks, and they are more efficient than sparse coding. An autoencoder is designed to reconstruct its inputs, which forces its middle (hidden) layer to form useful representations of the inputs. Then these representations can be used by a neural network for a supervised task such as classification. In other words, a stacked autoencoder learns something about the distribution of data and can be used to pre-train a neural network that has to operate on those data.

Therefore, many scientists contributed to the "invention" of deep learning and to the resurrection of neural networks. But the fundamental contribution came from Moore's Law: between the 1980s and 2006 computers had become enormously faster, cheaper and smaller. A.I. scientists were able to implement neural networks that were hundreds of times more complex, and able to train them with millions of data. This was still unthinkable in the 1980s. Therefore what truly happened between 1986 (when Restricted Boltzmann machines were invented) and 2006 (when deep learning matured) that shifted the balance from the logical approach to the connectionist approach in A.I. was Moore's Law. Without massive improvements in the speed and cost of computers deep learning would not have happened. Deep learning owes a huge debt of gratitude to the supercharged GPUs (Graphical Processing Units) that have become affordable in the 2010s.

Credit for the rapid progress in convolutional networks goes mainly to mathematicians, who were working on techniques for matrix-matrix multiplication and made their systems available as open-source software. The software, such as UC Berkeley's Caffe, used by neural-network designers, reduces a convolution to a matrix-matrix multiplication. This is a problem of linear algebra for which seasoned mathematicians had provided solutions. Initial progress took place at the Jet Propulsion Laboratory (JPL), a research center in California operated by the California Institute for Technology (CalTech) for the space agency NASA. Charles Lawson was the head of Applied Math Group at JPL since 1965. Lawson and his employee Richard Hanson developed software for linear algebra, including software for matrix computation, which was to be applied to astronomical things like gravitational fields. In 1979, together with Fred Krogh, an expert in differential equations, they released a Fortran library called Basic Linear Algebra Subprograms (BLAS). By 1990 BLAS 3 incorporated a library for matrix-matrix operations called General Matrix to Matrix Multiplication (GEMM), largely developed at Britain's Numerical Algorithms Group (instituted in 1970 as a joint project between several British universities and the Atlas Computer Laboratory). The computational "cost" of a neural network is mainly due to two kinds of layers: the layers that are fully-connected to each other and the convolutions. Both kinds entail massive multiplications of matrices; literally millions of them in the case of image recognition. Without something like GEMM no array of GPUs could perform the task.

A landmark achievement of deep-learning neural networks was published in 2012 by Alex Krizhevsky and Ilya Sutskever from Hinton's group at the University of Toronto: they demonstrated that deep learning (using a convolutional neural network with five convolutional layers and Bengio's rectified linear unit) outperforms traditional techniques of computer vision after processing 200 billion images during training (1.2 million human-tagged images plus thousands of computer-generated variants of each). Deep convolutional neural networks became the de facto standard for computer-vision systems.

In 2013 Google hired Hinton and Facebook hired LeCun.

Trivia: none of the protagonists of deep learning was born in the USA, although they all ended up working there. Fukushima is Japanese, LeCun and Bengio are French, Hinton is British, Ng is Chinese, Krizhevsky and Sutskever are Russian, Olshausen is Swiss. Add Hava Siegelmann from Israel, Sebastian Thrun and Sepp Hochreiter from Germany, Daniela Rus from Romania, Feifei Li from China, and the DeepMind founders from Britain and New Zealand.

Deep Belief Nets are probabilistic models that consist of multiple layers of probabilistic reasoning. Thomas Bayes' theorem of the

18th century is rapidly becoming one of the most influential scientific discoveries of all times (not bad for an unpublished manuscript discovered after Bayes' death). Bayes' theory of probability interprets knowledge as a set of probabilistic (not certain) statements and interprets learning as a process to refine those probabilities. As we acquire more evidence, we refine our beliefs. In 1996 the developmental psychologist Jenny Saffran showed that babies use probability theory to learn about the world, and they do learn very quickly a lot of facts. So Bayes had stumbled on an important fact about the way the brain works, not just a cute mathematical theory.

Since 2012 all the main software companies have invested in A.I. startups: Amazon (Kiva, 2012), Google (Neven, 2006; Industrial Robotics, Meka, Holomni, Bot & Dolly, DNNresearch, Schaft, Bost, DeepMind, Redwood Robotics, 2013-14), IBM (AlchemyAPI, 2015; plus the Watson project), Microsoft (Project Adam, 2014), Apple (Siri, 2011; Perceptio and VocalIQ, 2015; Emotient, 2016), Facebook (Face.com, 2012), Yahoo (LookFlow, 2013), Twitter (WhetLab, 2015), Salesforce (MetaMind, 2016), etc.

Since 2012 the applications of deep learning have multiplied. Deep learning has been applied to big data, biotech, finance, health care... Countless fields hope to automate the understanding and classification of data with deep learning.

Several platforms for deep learning have become available as open-source software: Torch (New York University), Caffe (Pieter Abbeel's group at UC Berkeley), Theano (Univ of Montreal, Canada), Chainer (Preferred Networks, Japan), Tensor Flow (Google), etc. This open-source software multiplies the number of people who can experiment with deep learning.

In 2015 Matthias Bethge's team at the University of Tübingen in Germany taught a neural network to capture an artistic style and then applied the artistic style to any picture.

The game of go/weichi had been a favorite field of research since the birth of deep learning. In 2006 Rémi Coulom introduced the Monte Carlo Tree Search algorithm and applied it to go/weichi. This algorithm dramatically improved the chances by machines to beat go masters: in 2009 Fuego Go (developed at the University of Alberta) beat Zhou Junxun, in 2010 MogoTW (developed by a French-Taiwanese team) beat Catalin Taranu, in 2012 Tencho no Igo/ Zen (developed by Yoji Ojima) beat Takemiya Masaki, in 2013 Crazy Stone (by Remi Coulom) beat Yoshio Ishida, and in 2016 AlphaGo (developed by Google's DeepMind) beat Lee Sedol. DeepMind's victory was widely advertised. DeepMind used a slightly modified Monte Carlo algorithm but, more importantly, it taught itself by playing against itself (a form of "reinforcement learning"). AlphaGo's neural network was trained with 150,000 games played by go/weichi masters. DeepMind had previously combined convolutional networks with reinforcement learning to train a neural network to play video games ("Playing Atari with Deep Reinforcement Learning", 2013).

Ironically, few people noticed that in September 2015 Matthew Lai unveiled an open-source chess engine called Giraffe that uses deep reinforcement learning to teach itself how to play chess (at international master level) in 72 hours. It was designed by just one person and it ran on the humble computer of his department at Imperial College London. (Lai was hired by Google DeepMind in January 2016, two months before AlphaGo's exploit against the Go master).

In 2016 Toyota demonstrated a self-teaching car, another application of deep reinforcement learning like AlphaGo: a number of cars are left to randomly roam the territory with the only rule that they have to avoid accidents. After a while, the cars learn how to drive properly in the streets.

The funny thing about convolutional networks is that nobody really knows why they work so well when they work well. Designing a convolutional network is still largely a process of "trial and error". In the paper "Why Does Deep And Cheap Learning Work So Well?" (2016) Henry Lin, a physicist at Harvard University, and Max Tegmark, a mathematician at the MIT, advanced the hypothesis that "deep learning" neural networks may have something profound in common with the nature of our universe.

The Robots are Coming – A Brief History of A.I./ Part 3

The story of robots is similar. Collapsing prices and increased speeds have enabled a generation of robots based on relatively old theory. Cynthia Breazeal's emotional robot "Kismet" (2000), Ipke Wachsmuth's conversational agent "Max" (2004), Honda's humanoid robot "Asimo" (2005), Osamu Hasegawa's robot that learned functions it was not programmed to do (2011) and Rodney Brooks' hand programmable robot "Baxter" (2012) look good on video but still look as primitive as Shakey in person. In 2005 the driver-less car Stanley developed by Sebastian Thrun at Stanford won DARPA's Grand Challenge, but that was in the middle of the Nevada desert. A branch of robotics is preoccupied with the self-reconfigurable modular robot, a concept introduced by Toshio Fukuda in Japan with its CEBOT (short for "cellular robot") that was capable of reconfiguring itself ("Self Organizing Robots Based On Cell Structures", 1988). The leadership remained in Japan (for example, Satoshi Murata's modular robotic system M-TRAN of 1999) until Daniela Rus at the MIT, inspired by the art of origami and a math theory by Erik Demaine, invented a robot that folds automatically ("Programmable Matter by Folding", 2010) which led to the self-configuring "M-blocks". Rus is also working on the Robot Compiler: someday we will be able to order a robot for a specific function and the Robot Compiler will 3D-print a custom robot for us.

Manufacturing plants have certainly progressed dramatically and can build, at a fraction of the cost, the tiny sensors and assorted devices that used to be unfeasible and that can make a huge difference in the movements of the robot; but there has been little conceptual breakthrough since Richard Fikes' and Nils Nilsson's STRIPS of 1969 (the "problem solver" used by Shakey). What is truly new is the techniques of advanced manufacturing and the speed of GPUs.

In fact, nothing puts the progress in A.I. (or lack thereof) better in perspective than the progress in robots. The first car was built in 1886. 47 years later (1933) there were 25 million cars in the USA, probably 40 million in the world, and those cars were much better than the first one. The first airplane took off in 1903. 47 years later (1950) 31 million people flew in airplanes, and those airplanes were much better than the first one. The first public radio broadcast took place in 1906. 47 years later, in 1953, there were more than 100 million radios in the world. The first television set was built in 1927. 47 years later (1974) 95% of households in the USA owned a TV set, and mostly a color TV set. The first commercial computer was delivered in 1951. 47 years later (1998) more than 40 million households in the USA had a computer, and those personal computers were more powerful than the first computer. The first (mobile) general-purpose robot was demonstrated in 1969 (Shakey). In 2016 (47 years later) how many people own a general-purpose robot? How many robots have you seen today in the streets or in your office?

In June 2016 the MIT Technology Review had an article about robots that announced: "They're invading consumer spaces including retail stores, hotels, and sidewalks". Look around you: how many robots do you see in the grocery shop and how many robots do you see taking a stroll on the sidewalk? I'll take a wild guess: zero. That's the great robot invasion of 2016, which competes with Orson Welles' famous Martian invasion of 1938 (total number of Martians in the streets of the USA: zero).

Most of the robots that accounted for the $28 billion market of 2015 (Tractica's estimate) were industrial robots, robots for the assembly line, not intelligent at all. Then there are more than ten million iRoomba (the home robot introduced by Rodney Brooks' iRobot in 2002) but those only vacuum floors. Those robots will never march in the streets to conquer Washington or Beijing. They are as intelligent as your washing machine, and not much more mobile.

Willow Garage, founded in 2006 by early Google architect Scott Hassan, has probably been the most influential laboratory of the last decade. They popularized the Robot Operating System (ROS), developed at Stanford in 2007, and they built the PR2 robot in 2010. ROS and PR2 have created a vast open-source community of robot developers that has greatly increased the speed at which a new robot can be designed. Willow Garage shut down in 2014, and its scientists founded a plethora of startups in the San Francisco Bay Area committed to developing "personal" robots.

The field of "genetic algorithms", or, better, evolutionary computing, has witnessed progress that mirrors the progress in neural-network algorithms; notably, in 2001 Nikolaus Hansen introduced the evolution strategy called "Covariance Matrix Adaptation" (CMA) for numerical optimization of non-linear problems. This has been widely applied to robotic applications and certainly helped better calibrate the movements of robots.

There are more than 3,000 DaVinci robots in the hospitals of the world, and they have performed about two million surgeries since 2000, the year when Intuitive Surgical of Sunnyvale was allowed to start deploying it. But DaVinci is only an assistant: it is physically operated by a human surgeon. In 2016, however, Peter Kim of the Children's National Health System in Washington unveiled a robot surgeon, the Smart Tissue Autonomous Robot (STAR), capable of performing an operation largely by itself (although it took about ten times longer than a human surgeon). In 2015 Google and Johnson & Johnson formed Verb Surgical to build robot surgeons.

The most sophisticated robots are actually airplanes. People rarely think of an airplane as a robot, but that's what it is: it mostly flies itself, from take-off to landing. In 2014 the world's airplanes carried 838.4 million passengers on more than 8.5 million flights. In 2015 a survey of Boeing 777 pilots reported that, in a typical flight, they spent just seven minutes manually piloting the airplane; and pilots operating Airbus planes spent half that time.

Therefore robots as "co-pilots" (as augmentation, not replacement, of human intelligence) have been very successful.

The most popular robot of 2016 is, instead, Google's self-driving car (designed by Sebastian Thrun), but this technology is at least 30 years old: Ernst Dickmanns demonstrated the robot car "VaMoRs" in 1986 and in October 1994 his modified Mercedes drove the Autoroute 1 near Paris in heavy traffic at speeds up to 130 km/h. In 2012 Google's co-founder Sergey Brin estimated that Google would have autonomous cars available for the general public within five years, i.e. by 2017. This is what happens when you think you know the future while in reality you don't even know the past. (Incidentally, Google engineers still use the "miles" of the ancient imperial system instead of the kilometers of the metric system, a fact that hardly qualifies as "progress" to me).

Intermezzo: Will Robots Be Conscious?

Whenever we discuss A.I. and robots, someone has to ask the question "Will these machines become conscious?"

The food industry slaughters 60 billion farmed animals (mammals, birds and fish) every year: why in heaven are we concerned for the consciousness of robots when we are not concerned for the consciousness of mammals and birds, whose brain is so similar to ours?

Brute-force A.I.

Despite all the hoopla, to me machines are still way less "intelligent" than most animals. Recent experiments with neural networks were hailed as sensational triumphs because a computer finally managed to recognize cats in videos (at least a few times) after being trained with 1.2 million labeled images. How long does it take for a mouse to learn how a cat looks like? And that's despite the fact that computers use the fastest possible communication technology, whereas the neurons of a mouse's brain use hopelessly old-fashioned chemical signaling.

One of the very first applications of neural networks was to recognize numbers. Sixty years later the ATM (automatic teller machine) of my bank still cannot recognize the amounts on many of the cheques that i deposit, but any human being can. Ray Kurzweil is often (incorrectly) credited with inventing "optical character recognition" (OCR), a technology that dates back to the 1950s (the first commercial OCR system was introduced by David Shepard's Intelligent Machines Research Corporation and became the basis for the Farrington Automatic Address Reading Machine delivered to the Post Office in 1953, and the term "OCR" itself was coined by IBM for its IBM 1418 product). Buy the most expensive OCR software and feed it the easiest possible case: a well-typed page from a book or magazine. It will probably make some mistakes that humans don't make, but, more interestingly, now slightly bend a corner of the page and try again: any human can still read the text, but the most sophisticated OCR software on the market will go berserk.

For similar reasons we still don't have machines that can read cursive handwriting, despite the fact that devices with handwriting recognition features already appeared in the 1990s (GO's PenPoint, Apple's Newton). Most people don't even know that their tablet or smartphone has such a feature: it is so inaccurate that very few people ever use it. And, yet, humans (even not very intelligent ones) can usually read other people's handwriting with little or no effort.

What has significantly improved is image recognition and speech recognition. Fei-fei Li's 2014 algorithm generates natural-language descriptions of images such as: "A group of men playing frisbee in a park". This result is based on a large dataset of images and their sentence descriptions that she started in 2009, ImageNet. In the 1980s it would have been computationally impossible to train a neural network with such a large dataset. The result may initially sound astounding (the machine algorithm even recognized the frisbee) but, even with the "brute force" of today's computers, in reality this is still a far cry from human performance: we easily recognize that those are young men, and many other details. And Peter Norvig of Google showed at Stanford's L.A.S.T. festival of 2015 a funny collection of images that were wrongly tagged by the machine because the machine has no common sense.

We are flooded with news of robots performing all sorts of human tasks, except that most of those tasks are useless. On the other hand, commenting on the ongoing unmanned Mars mission, in April 2013 NASA planetary scientist Chris McKay told me that "what Curiosity has done in 200 days a human field researcher could do in an easy afternoon." And that is the most advanced robotic explorer ever built.

What today's "deep learning" A.I. does is very simple: lots of number crunching. It is a smart way to manipulate large datasets for the purpose of classification. It was not enabled by a groundbreaking paradigm shift but simply by increased computing power.

The "Google Brain" project started at Google in 2011 by Andrew Ng (real name Wú Ēndá) is the quintessential example of this approach. In June 2012 a combined Google/Stanford research team used an array of 16,000 processors to create a neural network with more than one billion connections and let it loose on the Internet to learn from millions of YouTube videos how to recognize cats. Given the cost, size and speed of computers back then, 30 years ago nobody would have tried to build such a system. The difference between then and now is that today A.I. scientists can use thousands of powerful computers to get what they want. It is, ultimately, brute force with little or no sophistication. Whether this is how the human mind does it is debatable. And, again, we should be impressed that 16,000 of the fastest computers in the world took a few months to recognize a cat, something that a kitten with a still underdeveloped brain can do in a split second. I would be happy if the 16,000 computers could just simulate the 302-neuron brain of the roundworm, no more than 5000 synapses that nonetheless can recognize with incredible accuracy a lot of very interesting things.

The real innovation in Ng's approach was the idea to use GPUs. That simple idea made it possible to train multi-layer neural networks.

The human brain consumes about 20 Watts per hour. I estimate that AlphaGo's 1920 processors and 280 GPUs consumed about 440,000 Watts per hour (and that's not including the energy spent during the training process). What else can AlphaGo do besides playing Go? Absolutely nothing. What else can you do besides playing games? An infinite number of things, from cooking a meal to washing the car. AlphaGo consumed 440,000 W to do just one thing. Your brain uses 20 W and does an infinite number of things. How would you call someone that has to use 20,000 times more resources than you to do just one thing? What AlphaGo did is usually called "stupidity" not "intelligence". Let both the human and AlphaGo run on 20 Watts and see who wins. If it takes 440,000 Watts to play Go, how many Watts will it take to do everything else that a go/weichi master can do with his brain? Like driving a car, cooking a meal, jogging in the park, reading the news, chatting about literature with a friend, etc? A ridiculous number of machines will be needed to match the human capability, an amount of power perhaps exceeding the 15 terawatts that all nations combined consume. Perhaps it will take more machines that we can possibly build with all the materials available on the planet.

Brute force is the paradigm that now dominates A.I. After all, by indexing millions of webpages, a search engine is capable of providing an answer to the vast majority of questions (even "how to" questions), something that no expert system came close to achieving.

One wonders if slow and cumbersome computers were a blessing for the scientific community of the 1960s because those archaic machines forced computer scientists to come up with creative models instead of just letting modern high-speed computers crunch numbers until a solution is found.

John McCarthy was right to complain that, once A.I. solves a problem, the world does not consider it "artificial intelligence" anymore. But he didn't realize why: because, so far, whenever A.I. solved a problem (e.g., playing chess better than a master), the world realized that the solution wasn't special at all, it was just a matter of implementing very demanding mathematics on very fast computers. If 1+1=2 is not A.I., then playing chess is not A.I. A.I. has become synonymous with "running computationally intensive programs on supercomputers". We are impressed by the results, but we (correctly) don't consider them A.I. for the simple reason that human intelligence is something else. The name of the discipline is misleading. Not our fault.

Brute-force A.I. doesn't Speak

Most disheartening has been the scant progress in Natural Language Processing (NLP) since 1970.

Following are examples of exchanges between a person and Terry Winograd's program SHRDLU at the MIT in 1970:

Person: FIND A BLOCK WHICH IS TALLER THAN THE ONE YOU ARE HOLDING AND PUT IT INTO THE BOX.

Computer: BY "IT", i ASSUME YOU MEAN THE BLOCK WHICH IS TALLER THAN THE ONE i AM HOLDING.

Person: DOES THE SHORTEST THING THE TALLEST PYRAMID'S SUPPORT SUPPORTS SUPPORT ANYTHING GREEN?

Computer: YES, THE GREEN PYRAMID.

Person: WHAT COLOR IS IT?

Computer: BY "IT", i ASSUME YOU MEAN THE SHORTEST THING THE TALLEST PYRAMID'S SUPPORT SUPPORTS. RED

Person: DOES A STEEPLE

Computer interrupting: SORRY, i DON'T KNOW THE WORD "STEEPLE".

Person: A "STEEPLE" IS A STACK WHICH CONTAINS TWO GREEN CUBES AND A PYRAMID.

Computer: i UNDERSTAND.

Person: ARE THERE ANY STEEPLES NOW?
Computer: NO.

It beats some of today's "personal digital assistants".

We virtually abandoned the idea of having machines understand and speak our language and resorted to the opposite: make humans speak like machines. That is what you do when you talk on the phone with a machine that asks you for numbers; and that is what you do when you talk to your smartphone's "assistant" according to the rules of that assistant. Nothing illustrates the gap between machine intelligence and human intelligence better than comparing how much of a language a toddler learns in two years and how little of any language all machines ever invented have learned in more than 60 years.

IBM's Watson, that debuted with much fanfare in 2011 on a quiz show competing against human experts, was actually not capable of understanding the spoken questions: the questions were delivered to Watson as text files, not as spoken questions (a trick which, of course, distorted the whole game).

The most popular search engines are still keyword-based. Progress in search engines has been mainly in indexing and ranking webpages, not in understanding what the user is looking for nor in understanding what the webpage says. Try for example "Hey i had a discussion with a friend about whether Qaddafi wanted to get rid of the US dollar and he was killed because of that" and see what you get (as i write these words, Google returns first of all my own website with the exact words of that sentence and then a series of pages that discuss the assassination of the US ambassador in Libya). Communicating with a search engine is a far (far) cry from communicating with human beings.

Products that were originally marketed as able to understand natural language, such as SIRI for Apple's iPhone, have bitterly disappointed their users. These products understand only the most elementary of sounds, and only sometimes, just like their ancestors did decades ago. Promising that a device will be able to translate speech on the fly (like Samsung did with its Galaxy S4 in 2013) is a good way to embarrass yourself and to lose credibility among your customers.

During the 1960s, following Noam Chomsky's "Syntactic Structures" (1957) that heralded a veritable linguistic revolution, a lot of work in A.I. was directed towards "understanding" natural-language sentences, notably Charles Fillmore's case grammar at Ohio State University (1967), Roger Schank's conceptual dependency theory at Yale (1969) and William Woods' augmented transition networks at Harvard (1970). Unfortunately, the results were crude. Terry Winograd's SHRDLU and Woods' LUNAR (1973), both based on Woods' theories, were limited to very narrow domains and short sentences. We humans use all sorts of complicated sentences, some of them very long, some of them nested into each other. During the 1980s computer scientists such as Barbara Grosz at Harvard and Aravind Joshi at the University of Pennsylvania, as well as philosophers such as Hans Kamp in the Netherlands, attempted a more holistic approach to understanding "discourse", not just individual sentences; and this resulted in domain-independent systems such as the Core Language Engine, developed by Hiyan Alshawi's team at SRI in Britain. Meanwhile, Melvin Maron's pioneering work on statistical analysis of text was being resurrected by Gerard Salton at Cornell University (the project leader of SMART, System for the Mechanical Analysis and Retrieval of Text, since 1965). This technique represented a text as a "bag" of words, disregarding the order of the words and even the grammatical relationships. Surprisingly, this method was working better than the complex grammar-based approaches. It quickly came to be known as the "bag-of-words model" for language analysis. Technically speaking, it was text classification using naive Bayes classifiers. In 1998 Thorsten Joachims at Univ of Dortmund replaced the naive Bayes classifier with the method of statistical learning called "Support Vector Machines", invented by Vladimir Vapnik at Bell Labs in 1995, and other improvements followed. The bag-of-words model became the dominant paradigm for natural language processing but its statistical approach still failed to grasp the meaning of a sentence. In 2003 Yoshua Bengio at the University of Montreal used a different method for statistical language modeling, the kind of representation that is called "distributed" (as opposed to "local") in neural networks; and then in 2010 Andrew Ng at Stanford built on this mixed approach (neural networks and statistical analysis) using recursive neural networks.

The results are still far from human performance. The most illiterate person on the planet can understand language better than the most powerful machine.

Ironically, the biggest success story of the "bag-of-words" model has been in image classification, not in text classification. In 2003 Gabriela Csurka at Xerox in France applied the same statistical method to images. The "Bag-of-visual-words" model was born, that basically treats an image as a document. For the whole decade this was the dominant method for image recognition, especially when coupled with a Support Vector Machine classifier. This approach led, for example, to the system for classification of natural scenes developed in 2005 at Caltech by Pietro Perona and his student FeiFei Li at Caltech.

To be fair, progress in natural language understanding was hindered by the simple fact that humans prefer not to speak to another human in our time-consuming natural language. Sometimes we prefer to skip the "Good morning, how are you?" and get straight to the "Reset my Internet connection" in which case saying "One" to a machine is much more effective than having to wait for a human operator to pick up the phone and to understand your issue. Does anyone actually understand the garbled announcements in the New York subway? Communicating in natural language is not always a solution, as SIRI users are rapidly finding out on their smartphone. Like it or not, humans can more effectively go about their business using the language of machines. For a long time, therefore, Natural Language Processing remained an underfunded research project with few visible applications. It is only recently that interest in "virtual personal assistants" has resurrected the field.

Machine Translation too has disappointed. Despite recurring investments in the field by major companies, your favorite online translation system succeeds only with the simplest sentences, just like Systran in the 1970s. Here are some random Italian sentences from my old books translated into English by the most popular translation engine: "Graham Nash the content of which led nasal harmony", "On that album historian who gave the blues revival", "Started with a pompous hype on wave of hippie phenomenon".

The method that has indeed improved the quality of automatic translation is the statistical one, pioneered in the 1980s by Fred Jelinek's team at IBM and first implemented there by Peter Brown's team. When there are plenty of examples of (human-made) translations, the computer can perform a simple statistical analysis and pick the most likely translation. Note that the computer isn't even trying to understand the sentence: it has no clue whether the sentence is about cheese or parliamentary elections. It has "learned" that those few words in that combination are usually translated in such and such a way by humans. The statistical approach works wonders when there are thousands of (human-made) translations of a sentence, for example between Italian and English. It works awfully when there are fewer, like in the case of Chinese to English.

In 2013 Nal Kalchbrenner and Phil Blunsom of Oxford University attempted statistical machine translation based purely on neural networks ("Two Recurrent Continuous Translation Models"). In 2014 Ilya Sutskever's solved the "sequence-to-sequence problem" of deep learning using a Long Short-Term Memory ("Sequence to Sequence Learning with Neural Networks"), so the length of the input sequence of characters doesn't have to be the same length of the output.

Even if we ever get to the point that a machine can translate a complex sentence, here is the real test: "'Thou' is an Old English word". Translate that into Italian as "'Tu' e` un'antica parola Inglese" and you get an obviously false statement ("Tu" is not an English word). The trick is to understand what the original sentence means, not to just mechanically replace English words with Italian words. If you understand what it means, then you'll translate it as "'Thou' e` un'antica parola Inglese", i.e. you don't translate the "thou"; or, depending on the context, you might want to replace "thou" with an ancient Italian word like "'Ei' e` un'antica parola Italiana" (where "ei" actually means "he" but it plays a similar role to "thou" in the context of words that changed over the centuries). A machine will be able to get it right only when it fully understands the meaning and the purpose of the sentence, not just its structure.

(There is certainly at least one quality-assurance engineer who, informed of this passage in this book, will immediately enter a few lines of code in the machine translation program to correctly translate "'Thou' is an Old English word". That is precisely the dumb, brute-force, approach that i am talking about).

Or take Ronald Reagan's famous sarcastic statement, that the nine most terrifying words in the English language are "I'm from the government and i'm here to help". Translate this into Italian and you get "Le nove parole piu` terrificanti in Inglese sono `io lavoro per il governo e sono qui per aiutare'". Those are neither nine in the Italian translation (they are ten) and they are not "Inglese" (English) because they are now Italian. An appropriate translation would be "Le dieci parole piu` terrificanti in Italiano sono `io lavoro per il governo e sono qui per aiutare'". Otherwise the translation, while technically impeccable, makes no practical sense.

Or take Bertrand Russell's paradox: "the smallest positive integer number that cannot be described in fewer than fifteen words". This is a paradox because the sentence in quotes contains fourteen words. Therefore if such an integer number exists, it can be described by that sentence, which is fourteen words long. When you translate this paradox into Italian, you can't just translate fifteen with "quindici". You first need to count the number of words. The literal translation "il numero intero positivo piu` piccolo che non si possa descrivere in meno di quindici parole" does not state the same paradox because this Italian sentence contains sixteen words, not fourteen like the original English sentence. You need to understand the meaning of the sentence and then the nature of the paradox in order to produce an appropriate translation. I could continue with self-referential sentences (more and more convoluted ones) that can lead to trivial mistakes when translated "mechanically" without understanding what they are meant to do.

To paraphrase the physicist Max Tegmark, a good explanation is one that answers more than was asked. If i ask you "Do you know what time it is", a "Yes" is not a good answer. I expect you to at least tell me what time it is, even if it was not specifically asked. Better: if you know that i am in a hurry to catch a train, i expect you to calculate the odds of making it to the station in time and to tell me "It's too late, you won't make it" or "Run!" If i ask you "Where is the library?" and you know that the library is closed, i expect you to reply with not only the location but also the important information that it is currently closed (it is pointless to go there). If i ask you "How do i get to 330 Hayes St?" and you know that it used to be the location of a popular Indian restaurant that just shut down, i expect you to reply with a question "Are you looking for the Indian restaurant?" and not with a simple "It's that way". If i am in a foreign country and ask a simple question about buses or trains, i might get a lengthy lecture about how public transportation works, because the local people guess that I don't know how it works. Speaking a language is pointless if one doesn't understand what language is all about. A machine can easily be programmed to answer the question "Do you know what time it is" with the time (and not a simple "Yes"), and it can easily be programmed to answer similar questions with meaningful information; but we "consistently" do this for all questions, and not because someone told us to answer the former question with the time and other questions with meaningful information, but because that is what our intelligence does: we use our knowledge and common sense to formulate the answer.

In the near future it will still be extremely difficult to build machines that can understand the simplest of sentences. At the current rate of progress, it may take centuries before we have a machine that can have a conversation like the ones I have with my friends on the Singularity. And that would still be a far cry from what humans do: consistently provide an explanation that answers more than it was asked.

A lot more is involved than simply understanding a language. If people around me speak Chinese, they are not speaking to me. But if one says "Sir?" in English, and i am the only English speaker around, i am probably supposed to pay attention.

The state of Natural Language Processing is well represented by the results returned by the most advanced search engines: the vast majority of results are precisely the kind of commercial pages that i don't want to see. Which human would normally answer "do you want to buy perfume Katmandu" when i inquire about Katmandu's monuments? It is virtually impossible to find out which cities are connected by air to a given airport because the search engines all return hundreds of pages that offer "cheap" tickets to that airport.

Take, for example, zeroapp.email, a young startup being incubated in San Francisco in 2016. They want to use deep learning to automatically catalog the emails that you receive. Because you are a human being, you imagine that their software will read your email, understand the content, and then file it appropriately. If you were an A.I. scientist, you would have guessed instinctively that this cannot be the case. What they do is to study your behavior and learn what to do the next time that you receive an email that is similar to past ones. If you have done X for 100 emails of this kind, most likely you want to do X also for all the future emails of this kind. This kind of "natural language processing" does not understand the text: it analyzes statistically the past behavior of the user and then predicts what the user will want to do in the future. The same principle is used by Gmail's Priority Inbox, first introduced in 2010 and vastly improved over the years: these systems learn, first and foremost, by watching you; but what they learn is not the language that you speak.

I like to discuss with machine-intelligence fans a simple situation. Let's say you are accused of a murder you did not commit. How many years will it take before you are willing to accept a jury of 12 robots instead of 12 humans? Initially, this sounds like a question about "when will you trust robots to decide whether you are guilty or innocent?" but it actually isn't (i would probably trust a robot better than many of the jurors who are easily swayed by good looks, racial prejudices and many other unpredictable factors). The question is about understanding the infinite subtleties of legal debates, the language of lawyers and, of course, the language of the witnesses. The odds that those 12 robots fully understand what is going on at a trial will remain close to zero for a long time.

Can you hear me?

A brief summary of the field of speech recognition can serve to explain the infinite number of practical problems that must be solved in order to have a machine simply understand the words that i am saying (never mind the meaning of those words, just the words). A vast gulf separates popular books on the Singularity from the mundane daily research carried out at A.I. laboratories, where scientists work on narrow specialized technical details. Skip this chapter if you are bored by technical details, but trust me that the technical details are neither trivial nor few, and will keep several generations of engineers busy for a long time.

The history of speech recognition goes back at least to 1961, when IBM researchers developed the "Shoebox", a device that recognized spoken digits (0 to 9) and a handful of spoken words. In 1963 NEC of Japan developed a similar digit recognizer. Tom Martin at the RCA Laboratories was probably the first who applied neural networks to speech recognition ("Speech Recognition by Feature Abstraction Techniques", 1964). In 1970 Martin founded Threshold Technology in New Jersey which developed the first commercial speech-recognition product, the VIP-100.

Speech analysis became a viable technology thanks to conceptual innovations in Russia and Japan. In 1966 Fumitada Itakura at NTT in Tokyo invented Linear Predictive Coding ("One Consideration on Optimal Discrimination or Classification of Speech", 1966), a technique that 40 years later would be still used for voice compression in the GSM protocol for cellular phones; and Taras Vintsiuk at the Institute of Cybernetics in Kiev invented Dynamic Time Warping ("Speech Discrimination by Dynamic Programming", 1968), utilizing dynamic programming (a mathematical technique invented by Richard Bellman at RAND in 1953) to recognize words spoken at different speeds. Dynamic Time Warping was refined in 1970 by Hiroaki Sakoe and Seibi Chiba at NEC in Japan. Meanwhile, in 1969 Raj Reddy founded the speech-recognition group at Carnegie Mellon University and supervised three important projects: Harpy (Bruce Lowerre 1976), that used a finite-state network to reduce the computational complexity; Hearsay-II, that pioneered the "blackboard" in which knowledge acquired by parallel asynchronous processes gets integrated to produce higher levels of hypothesis (Rick Hayes-Roth, Lee Erman, Victor Lesser and Richard Fennell, 1975); and Dragon (Jim Baker, 1975), who moved to Massachusetts to start a pioneering company with the same name in 1982. Dragon differed from Hearsay in the way it represented knowledge: Hearsay used the logical approach of the "expert system" school, whereas Dragon used the Hidden Markov Model. The same idea was central to Fred Jelinek's efforts at IBM ("Continuous Speech Recognition by Statistical Methods", 1976) and these statistical methods based on the Hidden Markov Model for speech processing became popular with Jack Ferguson's "The Blue Book", which was the outcome of his lectures at the Institute for Defense Analyses in 1980.

IBM (Jelinek's group) and the Bell Labs (Lawrence Rabiner's group) came to represent two different schools of thought: IBM was looking for the individual speech-recognition system, that would be trained to recognize one specific voice; Bell Labs wanted a system that would understand a word pronounced by any one among the millions of AT&T's phone users. IBM studied the language model, whereas Bell Labs studied the acoustic model.

IBM's technology (the n-gram model) tried to optimize the recognition task by predicting statistically the next word. The inspiration for the IBM technique came from a word game devised by Claude Shannon in his book "A Mathematical Theory of Communication" (1948). Program this technique into a computer, and test it on your friends, and you have the Shannon equivalent of the Turing Test: ask both the computer and your friends to guess the next word in an arbitrary sentence. If the span of words is 1 or 2, your friends easily win. But if the span of words is 3 or higher, the computer starts winning.

Shannon's game was the first hint that perhaps understanding the meaning of the speech was irrelevant, and instead the frequency of each word and of its coexistence with other words was crucial.

Baum's Hidden Markov Model applied to speech recognition becomes a probability measure which integrates both schools because it can represent both the variability of speech sound and the structure of spoken language. The Bell Labs approach eventually led to Biing-Hwang Juang's "mixture-density hidden Markov models" for speaker independent recognition and a large vocabulary ("Maximum Likelihood Estimation for Mixture Multivariate Stochastic Observations of Markov Chains," 1985).

Hidden Markov Models became the backbone of the systems of the 1980s: Kai-Fu Lee's speaker-independent system Sphinx at Carnegie Mellon University (the most successful system yet for a large vocabulary and continuous speech); the Byblos system from BBN (1989); and the Decipher system from SRI (1989).

Three projects further accelerated progress in speech recognition. In 1989 Steve Young at Cambridge University developed the Hidden Markov Model Tool Kit, which soon became the most popular tool to build speech-recognition software. In 1991 Douglas Paul of the MIT, in collaboration with Dragon Systems, unveiled the Continuous Speech Recognition (CSR) Corpus, a dataset containing thousands of spoken articles, mostly from the Wall Street

Journal. Finally, in 1989 DARPA sponsored projects to develop speech recognition for air travel (the Air Travel Information Service or ATIS) with participants such as BBN, MIT, CMU, AT&T, SRI, etc. The program ended in 1994 when the yearly benchmark test showed that the error rate had dropped to human levels. These projects, largely based on Juang's algorithm of 1985, left behind another huge corpus of utterances. That's what you need to train speech-recognition systems. The following decade witnessed the first serious conversational agents: in 2000 Victor Zue at the MIT demonstrated Pegasus for airline flights status and Jupiter for weather status/forecast, and also in 2000 Al Gorin at AT&T developed How May I Help You (HMIHY) for telephone customer care. More importantly, the leader of the ATIS project at SRI, Michael Cohen, founded Nuance in 1994 that developed the system licensed by Siri to make the 2010 app for the Apple iPhone (and Cohen was hired by Google in 2004).

A Failed Experiment

In my opinion the "footnotes" in the history of Artificial Intelligence were not just footnotes: they represent colossal failures. They were all great ideas. In fact, they were probably the "right" ideas: of course, an intelligent machine must be capable of conversing in natural language; of course, it must be able to walk around, look for food, and protect itself; of course, it must be able to understand what people say (each person having a slightly different voice); of course, it would make more sense for software to "evolve" by itself than to be written by someone (just like any form of intelligent life did); of course, we would expect an intelligent machine to be able to write software (and build other machines, like we do); of course, it would be nice if the machine were capable of translating from one language to another; of course, it would make sense to build a computer that is a replica of a human brain if what we expect is a performance identical to the performance of a human brain.

These ideas have remained unfulfilled. In a sense, Artificial Intelligence is still a failed experiment: we still don't know how to do it properly.

Note that, ironically, it was A.I. that made computers popular and fueled progress in computer science. The idea of a thinking machine, not the usefulness of computers, drove the initial development. Since those days, progress in A.I. has been scant, but computers have become household appliances. Your laptop and smartphone are accidental by-products of a failed scientific experiment.

An Easy Science

When a physicist makes a claim, an entire community of physicists is out there to check that claim. The paper gets published only if it survives peer review, and usually many months after it was written. A discovery is usually accepted only if the experiment can be repeated elsewhere. For example, when OPERA announced particles traveling faster than light, the whole world conspired to disprove them, and eventually it succeeded. It took months of results before CERN accepted that probably (not certainly) the Higgs boson exists.

Artificial Intelligence practitioners, instead, have a much easier life. Whenever they announce a new achievement, it is largely taken at face value by the media and by the A.I. community at large. If a computer scientist announces that her or his program can recognize a cat, the whole world posts enthusiastic headlines even if nobody has actually seen this system in action, and nobody has been able to measure and doublecheck its performance: can i make a video of a cat and feed it into the program and see if it recognizes the cat? When in 2012 Google announced that "Our vehicles have now completed more than 300,000 miles of testing" (a mile being 1.6 kilometers for the rest of the world), the media simply propagated the headline without asking simple questions such as "in how many months?" or "under which conditions?" or "on which roads"? "at what time of the day"? Most people now believe that self-driving cars are feasible even though they have never been in one. Many of the same people probably don't believe all the weird consequences of Relativity and Quantum Mechanics, despite the many experiments that confirmed them.

The 2004 DARPA challenge for driverless cars was staged in the desert between Los Angeles and Las Vegas (i.e. with no traffic). The 2007 "DARPA Urban Challenge" took place at the George Air Force Base in California. Interestingly, a few months later two highly educated friends told me that a DARPA challenge took place in downtown Los Angeles in heavy traffic. That never took place. Too often the belief in the feats of A.I. systems feels like the stories of devout people who saw an apparition of a saint and all the evidence you can get is a blurred photo.

In 2005 the media reported that Hod Lipson at Cornell University had unveiled the first "self-assembling machine" (the same scientist in 2007 also unveiled the first "self-aware" robot), and in 2013 the media reported that the "M-blocks" developed at the MIT by Daniela Rus' team were self-constructing machines. Unfortunately, these reports were wild exaggerations.

In May 1997 the IBM supercomputer "Deep Blue", programmed by Feng-hsiung Hsu (who had started building chess-playing programs in 1985 while at Carnegie Mellon University), beat then chess world champion Garry Kasparov in a widely publicized match. What was less publicized is that the match was hardly fair: Deep Blue had been equipped with an enormous amount of information about Kasparov's chess playing, whereas Kasparov knew absolutely nothing of Deep Blue; and during the match IBM engineers kept tweaking Deep Blue with heuristics about Kasparov's moves. Even less publicized were the rematches, in which the IBM programmers were explicitly forbidden to modify the machine in between games. The new more powerful versions of Deep Blue (renamed Frintz) could beat neither Vladimir Kramnik, the new world chess champion, in 2002 nor Kasparov himself in 2003. Both matches ended in a draw. What is incredible to me is that a machine equipped with virtually an infinite knowledge of the game and of its opponent, and with lightning-speed circuits that can process virtually infinite number of moves in a split second cannot beat a much more rudimentary object such as the human brain equipped with a very limited and unreliable memory: what does it take for a machine to outperform humans despite all the technological advantages it has? Divine intervention? Nonetheless, virtually nobody in the scientific community (let alone in the mainstream media) questioned the claim that a machine had beaten the greatest chess player in the world.

If IBM is correct and, as it claimed at the time, Deep Blue could calculate 200 million positions per second whereas Kasparov's brain could only calculate three per second, who is smarter, the one who can become the world's champion with just three calculations per second or the one who needs 200 million calculations per second? If Deep Blue were conscious, it would be wondering "Wow, how can this human being be so intelligent?"

What Deep Blue certainly achieved was to get better at chess than its creators. But that is true of the medieval clock too, capable of keeping the time in a way that no human brain could, and of many other tools and machines.

Finding the most promising move in a game of chess is a lot easier than predicting the score of a Real Madrid vs Barcelona game, something that neither machines nor humans are even remotely close to achieving. The brute force of the fastest computers is enough to win a chess game, but the brute force of the fastest computers is not enough to get a better soccer prediction than, say, the prediction made by a drunk soccer fan in a pub. Ultimately what we are contemplating when a computer beats a chess master is still what amazed the public of the 1950s: the computer's ability to run many calculations at lightning speed, something that no human being can do.

IBM's Watson of 2013 consumes 85,000 Watts compared with the human brain's 20 Watts. (Again: let both the human and the machine run on 20 Watts and see who wins). For the televised match of 2011 with the human experts, Watson was equipped with 200 million pages of information including the whole of Wikipedia; and, in order to be fast, all that knowledge had to be stored on RAM, not on disk storage. The human experts who competed against Watson did not have access to all that information. Watson was allowed to store 15 petabytes of storage, whereas the humans were not allowed to browse the web or keep a database handy. De facto the human experts were not playing against one machine but against a whole army of machines, enough machines working to master and process all that data. A fairer match would be to pit Watson against thousands of human experts, chosen so as to have the same amount of data. And, again, the questions were conveniently provided to the machine as text files instead of spoken language. If you use the verb "to understand" the way we normally use it, Watson never "understood" a single question. And those were the easiest possible questions, designed specifically to be brief and unambiguous (unlike the many ambiguities hidden in ordinary human language). Watson didn't even hear the questions (they were written to it), let alone understand what the questioner was asking. Watson was allowed to ring the bell using a lightning-speed electrical signal, whereas the humans had to lift the finger and press the button, an action that is order of magnitudes slower.

Over the decades i have personally witnessed several demos of A.I. systems that required the audience to simply watch and listen: only the creator was allowed to operate the system.

Furthermore, some of the most headline-capturing Artificial Intelligence research is supported by philanthropists at private institutions with little or no oversight by academia.

Many of the A.I. systems of the past have never been used outside the lab that created them. Their use by the industry, in particular, has been virtually nil.

For example, on the first of October of 1999 *Science Daily* announced: "Machine demonstrates superhuman speech recognition abilities. University of Southern California biomedical engineers have created the world's first machine system that can recognize spoken words better than humans can". It was referring to a neural network trained by Theodore Berger's team. As far as i can tell, that project has been abandoned and it was never used in any practical application.

In October 2011 a Washington Post headline asked "Apple Siri: the next big revolution in how we interact with gadgets?" Meanwhile, this exchange was going viral on social media:

User: Siri, call me an ambulance
Siri: Okay, from now on I'll call you "an ambulance"

In 2014 the media announced that Vladimir Veselov's and Eugene Demchenko's program Eugene Goostman, which simulated a 13-year-old Ukrainian boy, passed the Turing Test at the Royal Society in London (Washington Post: "A computer just passed the Turing Test in landmark trial"). It makes you wonder what was the I.Q. of the members of the Royal Society, or, at least, of the event organizer, the self-appointed "world's first cyborg" Kevin Warwick, and what was the I.Q. of the journalists who reported his claims. It takes very little ingenuity to fool a "chatbot" impersonating a human being: "How many letters are in the word of the number that follows 4?" Any human being can calculate that 5 follows 4 and contains four letters, but a bot won't know what you are talking about. I can see the bot programmer, who has just read this sentence, frantically coding this question and its answer into the bot, but there are thousands, if not millions, of questions like this one that bots will fail for as long as they don't understand the context. How many words are in this sentence? You just counted them, right? But a bot won't understand the question. Of course, if your Turing Test consists in asking the machine questions whose answers can easily be found on Wikipedia by any idiot, then the machine will easily pass the test.

In 2015 both Microsoft and Baidu announced that their image-recognition software was outperforming humans, i.e. that the error rate of the machine was lower than the error rate of the average human being in recognizing objects. The average human error rate is considered to be 5.1%. However, Microsoft's technology that has surfaced (late 2015) is CaptionBot, which has become famous not for its usefulness in recognizing scenes but for the silly mistakes that no human being would make. As for Baidu, its Deep Image system, that ran on the custom-built supercomputer Minwa (432 core processors and 144 GPUs), has not been made available to the public as an app. However, Baidu was disqualified from the most prestigious image-recognition competition in the world (the ImageNet Competition) for cheating. Recognizing images was supposed to be Google's specialty but Google Goggles, introduced in 2010, has flopped. I just tried Goggles again (May 2016). It didn' recognize: towel, toilet paper, faucet, blue jeans... It recognized only one object: the clock. Officially, Google's image recognition software has an error rate of 5%. My test shows more like 90% error rate. In 2015 the Google Photos app tagged two African-Americans as gorillas, causing accusations of racism when in fact it was just poor technology. The media widely reported that Facebook's DeepFace (launched in 2015) correctly identified photos in 97.25% of cases (or so claimed Facebook), a fact that induced the European Union to warn Facebook that people's privacy must be protected; but in 2016 it identified none of my 5,000 friends: it works only if you have a small number of friends.

Sometimes the claims border on the ridiculous. Let's say that i build an app that asks you to submit the photo of an object, then the picture gets emailed to me, and i email back to you the name of the object: are you impressed by such an app? And, still, countless reviewers marveled at CamFind, the app introduced in 2013 by Los Angeles-based Image Searcher, an app that "recognizes" objects. In most cases it is actually not the app that recognizes objects, but their huge team in the Philippines that is frantically busy tagging the images submitted by users. Remember the automata of centuries ago, that in reality were people camouflaged like machines? In 1769, a chess-playing machine called "The Turk", created by Wolfgang von Kempelen, toured the world, winning games wherever it went: it concealed a man inside so well that it wasn't exposed as a hoax for many years.

(To be fair, Microsoft's CaptionBot is not bad at all: it was criticized by people who expected human-level abilities in the machine, but, realistically, it exceeds my expectations).

Very few people bother to doublecheck the claims of the A.I. community. The media have a vested interest that the story be told (it sells) and the community as a whole has a vested interest that government and donors believe in the discipline's progress so that more funds are poured into it.

Paul Nunez in "Brain, Mind, and the Structure of Reality" (2010) distinguishes between Type 1 scientific experiments and Type 2 experiments. Type 1 is an experiment that has been repeated at different locations by different teams and still holds. Type 2 is an experiment that has yielded conflicting results at different laboratories. UFO sightings, levitation tales and exorcisms are not scientific, but many people believe in their claims, and i will call them "Type 3" experiments, experiments that cannot be repeated by other scientists. Too much of Artificial Intelligence occupies the space between Type 2 and Type 3.

News of the feats achieved by machines rapidly propagate worldwide thanks to enthusiastic bloggers and tweeters the same way that news about telepathy and levitation used to spread rapidly worldwide thanks to word of mouth without the slightest requirement of proof. (There are still millions of people who believe that cases of levitation have been documented even though there is no footage and no witness to be found anywhere). The belief in miracles worked the same way: people wanted to believe that a saint had performed a miracle and they transmitted the news to all their acquaintances in a state of delirious fervor without bothering to doublecheck the facts and without providing any means to doublecheck the facts (address? date? who was there? what exactly happened?). The Internet is a much more powerful tool than the old "word of mouth" system. In fact, i believe that part of this discussion about machine intelligence is a discussion not about technology but about the World-wide Web as the most powerful tool ever invented to spread myths. And part of this discussion about machine intelligence is a discussion about the fact that 21st century humans want to believe that super-intelligent machines are coming the same way that people of previous centuries wanted to believe that magicians existed. The number of people whose infirmity has been healed after a visit to the sanctuary of Lourdes is very small (and in all cases one can find a simple medical explanation) but thousands of highly educated people still visit it when they get sick, poor or depressed. On 13 October 1917 in Fatima (Portugal) tens of thousands of people assembled because three shepherd children had been told by the Virgin Mary (the mother of Jesus) that she would appear at high noon. Nobody saw anything special (other than the Sun coming out after a rain) but the word that a miracle had taken place in Fatima spread worldwide. Believe it or not, that is pretty much what happened in 2013 when a fanatic blogger reported a feat performed by an A.I. software or a robot as a new step towards the Singularity. People like me who remain skeptical of the news are frowned upon in the same way that skeptics were looked down upon after Fatima: "What? You still don't believe that the Virgin Mary appeared to those children? What is wrong with you?". Which, of course, shifts the burden of proof on the skeptic who is asked to explain why one would NOT believe in the miracle (sorry, i meant "in the machine's intelligence") instead of pressing the inventor/scientist/lab/firm into proving that the miracle/feat has truly been accomplished and can be repeated at will and that it really did what bloggers said it did.

   "Whenever a new science achieves its first big successes, its enthusiastic acolytes always fancy that all questions are now soluble" (Gilbert Ryle, "The Concept of Mind", 1949, six years before Artificial Intelligence was born).

Intermezzo and Trivia: the Original App

At the same time the real achievements of the machine are sometime neglected. I am not terribly impressed that computers can play chess. I am much more impressed that computers can forecast the weather, since the atmosphere is a much more complex system than the game of chess. The media have always devoted more attention to the game of chess because its rules are easier to explain to the general public, whereas the rules that guide air flow and turbulence are rather exotic. However, it turns out that weather forecasting was the original "app".

Weather forecast was the "mission impossible" of the early computers. The first weather forecast using a computer simulation dates back to March 1950, to the early history of electronic computers. The computer was an ENIAC and it took just about 24 hours to calculate the weather forecast for the next 24 hours. Weather forecasting was a particularly interesting application of electronic computing for John Von Neumann. In fact, it was "the" application originally envisioned for the machine that Von Neumann designed at Princeton's Institute for Advanced Studies (IAS), the machine that introduced the "Von Neumann architecture" still used today. Mathematicians had known for a while that solving this problem, i.e. modeling the air flow, required solving a non-linear system of partial differential equations - Lewis Richardson had published the milestone study in this field, "Weather Prediction by Numerical Process" in 1922 - and that is why mathematicians thought this was an avantgarde problem; and that's why Von Neumann felt that solving it with a computer would not only help the community of meteorologists but also prove that the electronic computer was no toy. The ENIAC program, however, used an approximation devised by Jule Charney in 1948 ("On the scale of atmospheric motions"). A computer model for the general circulation of the atmosphere had to wait until 1955, when Norman Phillips, also at Princeton, presented his equations at the Royal Meteorological Society, and fed them into the IAS computer Maniac i ("The general circulation of the atmosphere", 1955). Meanwhile, the ability to predict the weather was dramatically improved in 1957 when the first satellite was launched. By 1963 a Japanese scientist at UCLA, Akio Arakawa, had tweaked Phillips' equations and written a Fortran program on an IBM 709, with help from IBM's Large Scale Scientific Computation Department in San Jose ("Computational Design for Long-Term Numerical Integration of the Equations of Fluid Motion", 1966). IBM was obviously ecstatic that their computer could be used to solve such a strategic problem as predicting the weather. It was the Fortran programming language's baptism of fire, as the 709 was the first commercial computer equipped with a Fortran compiler. At this point it looked like it was just a matter of waiting for computers to get faster. Alas, in the same year that Arakawa produced the first meaningful weather forecast, Edward Lorenz proved that the atmosphere belongs to the class of systems now known as "chaotic" ("Deterministic Nonperiodic Flow", 1963): there is a limit to how accurately one can predict their behavior. In fact, as computers grow exponentially faster due to Moore's law, weather forecast models have not become exponentially more accurate. Robin Stewart has shown that "despite this exponential increase in computational power, the accuracy of forecasts has increased in a decidedly linear fashion" ("Weather Forecasting by Computers", 2003). Even today meteorologists can only give us useful forecasts of up to about a week.

Note that, unlike chess and machine translation, this problem is not currently solved by using statistical analysis. It is solved by observing the current conditions and applying physical laws (as derived by those pioneering scientists). Statistical analysis requires an adequate sample of data, and a relatively linear behavior. Weather conditions, instead, are never the same, and the nonlinear nature of chaotic systems like the atmosphere makes it very easy to come up with grotesquely wrong predictions. This does not mean that it is impossible to predict the weather using statistical analysis; just that it is only one method out of many, a method that has been particularly successful in those fields where statistical analysis makes sense but was not feasible before the introduction of powerful computers. There is nothing magical about its success, just like there is nothing magical about our success in predicting the weather. Both are based on good old-fashioned techniques of computational mathematics.

Don't be Fooled by the Robot

The bar is being set very low for robotics too. Basically, any remote-controlled toy (as intelligent as the miniature trains that were popular in the 1960s) is now being hailed as a step toward the robot invasion. I always advise robotics fans to visit the Musee Mecanique in San Francisco, that has a splendid collection of antique coin-operated automatic mechanical musical instruments... sorry, i meant "robotic musicians", before they venture into a discussion about progress in robotics. These automata don't constitute what we normally call "intelligence" but they provide amazing shows. Automata have entertained royalties and peasants for centuries: Ismail Al-Jazari's music ensemble of 1206, Leonardo DaVinci's knight of 1495, Juanelo Turriano's monk of 1560, Jacques de Vaucanson's duck of 1739, Pierre Jaquet-Droz's dolls of 1768-74, John Joseph Merlin's "Silver Swan" of 1773, Hubert Martinet's musical elephant of 1774, Henri Maillardet's draughtsman-writer of 1800, Joseph Faber's Euphonia of 1840... the list is endless.

Does driving a car qualify as a sign of "intelligence"? Maybe it does, but it has to be "really" what it means for humans. There is no car that has driven even one meter without help from humans. The real world is a world in which first you open the garage door, then you stop to pick up the newspaper, then you enter the street and you will stop if you see a pedestrian waiting to cross the street. No car has achieved this skill yet. They self-drive only in highly favorable conditions on well marked roads with well marked lanes and only on roads that the manufacturing company has mapped accurately (in other words, with a lot of help from humans). And i will let you imagine what happens if the battery dies or there's a software bug... What does the self-driving car do if it is about to enter a bridge when an earthquake causes the bridge to collapse? Presumably it will just drive on. What does the self-driving car do if it is stopping at a red light and a man with a gun breaks the window? Probably nothing: it's a red light. If you fall asleep in a self-driving car, your chances of dying will skyrocket. There are countless rules of thumb that a human driver employs all the time, and they are based on understanding what is going on. A set of sensors wrapped in a car's body does not understand anything about what is going on.

Human-looking automata that mimic human behavior have been built since ancient times and some of them could perform sophisticated movements. They were mechanical. Today we have electromechanical sophisticated toys that can do all sort of things. There is a (miniature) toy that looks like a robot riding a bicycle. Technically speaking, the whole toy is the "robot". Philosophically speaking, there is no robot riding a bicycle. The robot-like thing on top of the bicycle is redundant, it is there just for show: you can remove the android and put the same gears in the bicycle seat or in the bicycle pedals and the bike with no passenger would go around and balance itself the exact same way: the thing that rides the bicycle is not the thing on top of the bike (designed to trick the human eye) but the gear that can be placed anywhere on the bike. The toy is one piece: instead of one robot, you could put ten robots on top of each other, or no robot at all. Any modern toy store has toys that behave like robots doing some amazing thing (amazing for a robot, ordinary for a human). It doesn't require intelligence: just good engineering. This bike-riding toy never falls, even when it is not moving. It is designed to always stand vertical. Or, better, it falls when it runs out of battery. That's very old technology. If that's what we mean by "intelligent machines", then they have been around for a long time. We even have a machine that flies in the sky using that technology. Does that toy represent a quantum leap forward in intelligence? Of course, no. It is remotely controlled just like a television set. It never "learned" how to bike. It was designed to bike. And that's the only thing it can do. The only thing that is truly amazing in these toys is the miniaturization, not the "intelligence".

If you want this toy to do something else, you'll have to add more gears of a different kind, specialized in doing that other thing. Maybe it is possible (using existing technology or even very old mechanical technology) to build radio-controlled automata that have one million different gears to do every single thing that humans do, the whole taking up no more space than my body does. It would still be a toy.

A human being is NOT a toy (as yet).

The Curse of the Large Dataset

The most damning evidence that A.I. has posted very little conceptual progress towards human-level intelligence comes from an analysis of what truly contributed to Deep Learning's most advertised successes of recent years: the algorithm or the training database? The algorithm is designed to learn an intelligent task, but it has to be trained via human-provided examples of that intelligent task.

There is a pattern about neural networks that has become the norm after the 1990s: an old technique stages spectacular performance thanks to a large training dataset, besides more powerful processors.

In 1997 Deep Blue used Reinefeld's NegaScout algorithm of 1983. The key to its success, besides the massively-parallel 30 high-performance processors, was a dataset of 700,000 chess games played by masters, a dataset created in 1991 by IBM for the second of Feng-hsiung Hsu's chess-playing programs, Deep Thoughts 2.

In 2011 Watson utilized (quote) "90 clustered IBM Power 750 servers with 32 Power7 cores running at 3.55 GHz with four threads per core" and a dataset of 8.6 million documents culled from the Web in 2010, but its "intelligence" was Robert Jacobs' 20-year-old "mixture-of-experts" technique.

All the successes of convolutional neural networks after 2012 were based on Fukushima's 30-year-old technique but trained on the ImageNet dataset of one million labeled images created in 2009 by Feifei Li.

In 2015 DeepMind's videogame-playing program used Chris Watkins' Q-learning algorithm of 1989 but trained on the Arcade Learning Environment dataset of Atari games developed in 2013 by Michael Bowling's team at the University of Alberta.

It is easy to predict that the next breakthrough in Deep Learning will not come from a new conceptual discovery but from a new large dataset in some other domain of expertise. Progress in Deep Learning depends to a large extent on many human beings (typically PhD students) who manually accumulate a large body of facts. It is not terribly important what kind of neural network gets trained to use those data, as long as there are really a lot of data. The pattern looks like this: at first the dataset becomes very popular among A.I. hackers; then some of these hackers utilize an old-fashioned A.I. technique to train a neural network until it exhibits master-like skills in that domain.

A Consumer's Rant Against the Stupidity of Machines: Reversing Evolution?

When you buy an appliance and it turns out that you have to do something weird in order to make it work, it is natural to dismiss it as "a piece of garbage". However, when it is something about computers and networks, you are instead supposed to stand in awe and respectfully listen to (or read) a lengthy explanation of what you are supposed to do in order to please the machine, which is usually something utterly convoluted bordering on the ridiculous.

This double standard creates the illusion that machines are becoming incredibly smart when in fact mostly we are simply witnessing poor quality assurance (due to the frantic product lifecycles of our times) and often incredibly dumb design.

You never know what is going to happen to your favorite application when you download an "update". New releases (which you are forced to adopt even if you are perfectly happy with the old release) often result in lengthy detours trying to figure out how to do things that were trivial in the previous release (and that have been complicated by the software manufacturer for no other reason than to justify a new release). A few weeks ago my computer displayed the message "Updating Skype... Just a moment, we're improving your Skype experience". How in heaven do they know that this will improve my Skype experience? Of course they don't. The reason they want you to move to a new release is different: it will certainly improve THEIR experience. Whether it will also improve mine and yours is a secondary issue. At the least, any change in the user interface will make it more difficult to do the things to which you were accustomed.

We live in an age in which installing a wireless modem can take a whole day and external hard disks get corrupted after a few months "if you use them too often" (as a sales associate told me at Silicon Valley's most popular computer store).

In the old days i was backing up my work all the time because i didn't trust computers: they frequently crashed. I didn't trust them because they were not reliable. These days computers don't crash anymore but i still back up frequently my work: i don't trust them because it is unpredictable what they will do with my work. Computers download and store files where they want (often in obscure folders/ directories), sometimes they change my desktop appearance, sometimes they change the formatting of my documents, etc. The manufacturers that sell them tell me that these programs are becoming "more and more intelligent": obviously we have wildly different definitions of "intelligence".

In 1997 Steve Jobs famously told Business Week: "People don't know what they want until you show it to them." Maybe. But sometimes the high-tech industry should say: "People don't know what they want until we FORCE it upon them and give them no alternative."

Reality check: here is the transcript of a conversation with Comcast's automated customer support:

"If you are currently a Comcast customer, press 1" [I press 1]

"Please enter the ten-digit phone number associated with your account" [I enter my phone number]

"OK Please wait just a moment while i access your account"

"For technical help press 1"

"For billing press 2" [I press 2]

If you are calling regarding important information about Xfinity etc press 1 [I press 2]

"For payments press 1"

"For balance information press 2"

"For payment locations press 3"

"For all other billing questions press 4" [I press 4]

"For questions about your first bill press 1"

"For other billing questions press 3" [I press 3]

"Thank you for calling Comcast. Our office is currently closed."

(You can listen to it at https://soundcloud.com/scaruffi/comcast-customer-support)

Based on the evidence, it is easier to believe that we still live in the stone age of computer science than to believe that we are about to witness the advent of superhuman intelligence in machines.

It is interesting how different generations react to the stupidity of machines: the old generation that grew up without electronic machines gets extremely upset (because the automated system can complicate things that used to be simple in the old-fashioned manual system), my generation (that grew up with machines) gets somewhat upset (because machines are still so dumb), and the younger generations are progressively less upset, with the youngest ones simply taking for granted that customer support has to be what it is (from lousy to non-existent) and that many things (pretty much all the things that require common sense, expertise, and what we normally call "intelligence") are virtually impossible for machines.

A book on "The State of Machine Stupidity" instead of "The State of Machine Intelligence" should be much longer.

Incidentally, there are very important fields, such as getting rid of paper, in which we haven't even achieved the first step of automation. Health care, for example, still depends on paper: your medical records are probably stored in old fashioned files, not the files made of zeros and ones but the ones made of cardboard or plastic. We are bombarded daily by news of amazing medical devices and applications that will change the way diseases are prevented, identified and healed, but for the time being we have seen very little progress in simply turning all those paper files into computer files that the patient can access from a regular computer or smartphone and then save, print, email or delete at will.

What we have done so far, and only in some fields, is to replace bits and pieces of human intelligence with rather unintelligent machines that can only understand very simple commands (less than what a two-year old toddler can understand) and can perform very simple tasks.

In the process we are also achieving lower and lower forms of human intelligence, addicted to having technology simplify all sorts of tasks (more about this later). Of course, many people claim the opposite: from the point of view of a lower intelligence, what unintelligent machines do might appear intelligent.

The Singularity as the Outcome of Exponential Progress

The Singularity crowd is driven to enthusiastic prognostications about the evolution of machines: machines will soon become intelligent and they will rapidly become intelligent in a superhuman way, acquiring a higher form of intelligence than human intelligence.

There is an obvious disconnect between the state of the art and what the Singularity crowd predicts. We are not even remotely close to a machine that can troubleshoot and fix an electrical outage or simply your washing machine, let alone a software bug. We are not even remotely close to a machine that can operate any of today's complex systems without human supervision. One of the premises of the theory of the Singularity is that machines will not only become intelligent but will even build other, smarter machines by themselves; but right now we don't even have software that can write other software.

The jobs that have been automated are repetitive and trivial. And in most cases the automation of those jobs has required the user/customer to accept a lower (not higher) quality of service. Witness how customer support is rapidly being reduced to a "good luck with your product" kind of service. The more the automation around you, the more you are forced to behave like a machine in order to interact and communicate with machines, precisely because they are still so dumb.

The reason that we have a lot of automation is that (in developed countries like the USA, Japan and the European countries) it saves money: machine labor is a lot cheaper than human labor. Wherever the opposite is true, there are no machines. The reason we are moving to online education is not that university professors failed to educate their students but that universities are too expensive. And so forth: in most cases it is the business plan, not the intelligence of machines, that drives automation.

Wildly optimistic predictions are based on the exponential progress in the speed and miniaturization of computers. In 1965 Gordon Moore predicted that the processing power of computers would double every 18 months ("Moore's law"), and so far his prediction has been correct. Look closer and there is little in what they say that has to do with software. It is mostly a hardware argument. And that is not surprising: predictions about the future of computers have been astronomically wrong in both directions but, in general, the ones that were too conservative were about hardware (its progress has surprised us), the ones that were too optimistic were about software (its progress has disappointed us). What is amazing about today's smartphones is not that they can do what computers of the 1960s could not do, but that they are small, cheap and fast. The fact that there are many more software applications downloadable for a few cents means that many more people can use them, a fact that has huge sociological consequences; but it does not mean that a conceptual breakthrough has been reached in software technology. It is hard to name one software program that exists today and could not have been written in Fortran fifty years ago. If it wasn't written, the reason, probably, is that it would have been too expensive or that some required hardware did not exist yet.

Accelerating technological progress in computer science has largely been driven by the accelerating cost of labor, not by real scientific innovation. The higher labor costs go, the stronger the motivation to develop "smarter" machines. Those machines, and the underlying technologies, were already feasible ten or twenty or even thirty years ago, but back then it didn't make economic sense for them to be adopted.

There has certainly been a lot of progress in computers getting faster, smaller and cheaper. Even assuming that this will continue "exponentially" (as the Singularity crowd is quick to claim), the argument that this kind of (hardware) progress is enough to make a shocking difference in terms of machine intelligence is based on an indirect assumption: that faster/smaller/cheaper will lead first to a human-level intelligence and then to a superior intelligence. After all, if you join together many many many dumb neurons you get the very intelligent brain of Albert Einstein. If one puts together millions of superfast GPUs, maybe one gets superhuman intelligence. Maybe.

In any event, we'd better prepare for the day that Moore's Law stops working. Moore's Law was widely predicted to continue in the foreseeable future, but its future does not look so promising anymore. It is not only that technology might be approaching the limits of its capabilities, but also the original spirit behind Moore's Law was to show that the "cost" of making transistor-based devices would continue to decline. Even if the industry finds a way to continue to double the number of transistors etched on a chip, the cost of doing so might start increasing soon: the technologies to deal with microscopic transistors are inherently expensive, and heat has become the main problem to solve in ultra-dense circuits. In 2016 William Holt of Intel announced that Intel will not push beyond the 7-nanometer technology, and cautioned that processors may get slower in the future in order to save energy and reduce heat, i.e. costs. For 70 years computers have been getting smaller and smaller, but in 2014 they started getting bigger again (the iPhone 6 generation). If Moore's Law stops working, will there still be progress in "Brute-force A.I.", e.g. in deep learning? In 2016 Scott Phoenix, the CEO of Silicon Valley-based AI startup Vicarious, declared that "In 15 years, the fastest computer will do more operations per second than all the neurons in all the brains of all the people who are alive." What if this does not come true?

The discussion about the Singularity is predicated upon the premise that machines will soon be able to perform "cognitive" tasks that were previously exclusive to humans. This, however, has already happened. We just got used to it. The early computers of the 1950s were capable of computations that traditionally only the smartest and fastest mathematicians could even think of tackling, and the computers quickly became millions of times faster than the fastest mathematician. If computing is not an "exclusively human cognitive task", i don't know what would qualify. Since then computers have been programmed to perform many more of the tasks that used to be exclusive to human brains. And no human expert can doublecheck in a reasonable amount of time what the machine has computed. Therefore there is nothing new about a machine performing a "cognitive" task that humans cannot match. Either the Singularity already happened in the 1950s or it is not clear what cognitive task would represent the coming of the Singularity.

To assess the progress in machine intelligence one has to show something (some intelligent task) that computers can do today that, given the same data, they could not have done fifty years ago.

There has been a lot of progress in miniaturization and cost reduction, so that today it has become feasible to use computers for tasks for which we didn't use them fifty years ago; not because they were not intelligent enough to do them but because it would have been too expensive and it would have required several square kilometers of space. If that's "artificial intelligence", then we invented artificial intelligence in 1946. Today's computers can do a lot more things than the old ones just like new models of any machine (from kitchen appliances to mechanical reapers) can do a lot more things than old models. Incremental engineering steps lead to more and more advanced models for lower prices. Some day a company will introduce coffee machines on wheels that can make the coffee and deliver the cup of coffee to your desk. And the next model will include voice recognition that understands "coffee please". Etc. This kind of progress has been going on since the invention of the first mechanical tool. It takes decades and sometimes centuries for the human race to fully take advantage of a new technology. "Progress" often means the process of mastering a new technology (of creating ever more sophisticated products based on that technology). The iPhone was not the first smartphone, and Google was not the first search engine, but we correctly consider them "progress".

There is no question that progress has accelerated with the advent of electrical tools and further accelerated with the invention of computers. Whether these new classes of artifacts eventually constitute a different kind of "intelligence" will probably depend on your definition of "intelligence".

The way the Singularity would be achieved by intelligent machines is by these machines building more intelligent machines capable of building more intelligent machines and so forth. A similar loop has existed since about 1776. The steam engine enabled mass production of steel, which in turn enabled the mass production of better steam engines, and this recursive loop continued for a while. James Watt himself, inventor of the steam engine that revolutionized the world, worked closely with John

Wilkinson, who made the steel for Watt's engines using Watt's engines to make the steel. Today this loop of machines helping build other machines takes place on a large scale. For example, a truck carries the materials that the factory will use to make better trucks. The human beings in this process can be viewed as mere intermediaries between machines that are evolving into better machines. This positive-feedback loop is neither new nor necessarily "exponential". In the 19th century that loop of machines building (better) machines which build (better) machines accelerated for a while. Eventually, the steam engine (no matter how sophisticated that accelerating positive-feedback loop had made it) was made obsolete by a new kind of machine, the electrical motor. Again, electrical motors were used by manufacturers of motor parts that contributed to making better electrical motors used by manufacturers of electrical motor parts.

We have been surrounded by machines that built better machines for a long time... but with human intermediaries designing the improvements.

Despite the fact that no machine has ever created another machine of its own will, and no software has ever created a software program of its own will, the Singularity crowd seems to have no doubts that a machine is coming soon created by a machine created by a machine and so forth, each generation of machines being smarter than the previous one.

i certainly share the concern that the complexity of a mostly automated world could get out of hand. This concern has nothing to do with the degree of intelligence but just with the difficulty of managing complex systems. Complex, self-replicating systems that are difficult to manage have always existed. For example: cities, armies, post offices, subways, airports, sewers, economies...

"Machines are, or are becoming, animate" (Samuel Butler in 1863, when not even the light bulb had been invented)

A Look at the Evidence: A Comparative History of Accelerating Progress

A postulate at the basis of many contemporary books by futurists and self-congratulating technologists is that we live in an age of unprecedented rapid change and progress. But look closer and our age won't look so unique anymore.

As i wrote in the chapter titled "Regress" of my book "Synthesis", this perception that we live in an age of rapid progress is mostly based on the fact that we know the present much better than we know the past. One century ago, within a relatively short period of time, the world adopted the car, the airplane, the telephone, the radio and the record, while at the same time the visual arts went through Impressionism, Cubism and Expressionism. Science was revolutionized by Quantum Mechanics and Relativity. Office machines (cash registers, adding machines, typewriters) and electrical appliances (dishwasher, refrigerator, air conditioning) dramatically changed the way people worked and lived. Debussy, Schoenberg, Stravinsky and Varese changed the concept of music. These all happened in one generation. By comparison, the years since World War II have witnessed innovation that has been mostly gradual and incremental. We still drive cars (invented in 1886) and make phone calls (invented in 1876), we still fly on airplanes (invented in 1903) and use washing machines (invented in 1908), etc. Cars still have four wheels and planes still have two wings. We still listen to the radio and watch television. While the computer and Genetics have introduced powerful new concepts, and computers have certainly changed daily lives, i wonder if any of these "changes" compare with the notion of humans flying in the sky and of humans located in different cities talking to each other. There has been rapid and dramatic change before.

Does the revolution in computer science compare with the revolutions in electricity of a century ago? The smartphone and the Web have certainly changed the lives of millions of people, but didn't the light bulb, the phonograph, the radio and kitchen appliances change the world at least as much if not much more?

A history of private life in the last 50 years would be fairly disappointing: we wear pretty much the same clothes (notably Tshirts and blue jeans), listen to the same music (rock and soul were invented in the 1950s), run in the same shoes (sneakers date from the 1920s), and ride, drive and fly in the same kinds of vehicles (yes, even electric ones: Detroit Electric began manufacturing electric cars in 1907). Public transportation is still pretty much what it was a century ago: trams, buses, trains, subways. New types of transportation have been rare and have not spread widely: the monorail (that became reality with the Tokyo Monorail in 1964), the supersonic airplane (the Concorde debuted in 1976 but was retired in 2003), the magnetic levitation train (the Birmingham Maglev debuted in 1984, followed by Berlin's M-Bahn in 1991, but in practice the Shanghai Maglev Train built in 2004 is the only real high-speed magnetic levitation line in service). The "bullet train" (widely available in Western Europe and the Far East since Japan's Shinkansen of 1964) is probably the only means of transportation that has significantly increased the speed at which people travel long distances in the last 50 years.

We chronically underestimate progress in previous centuries because most of us are ignorant about those eras. Historians, however, can point at the spectacular progress that took place in Europe during the Golden Century (the 13th century) when novelties such as spectacles, the hourglass, the cannon, the loom, the blast furnace, paper, the mechanical clock, the compass, the watermill, the trebuchet and the stirrup changed the lives of millions of people within a few generations; or the late 15th century when (among other things) the printing press enabled an explosive multiplication of books and when long-distance voyages to America and Asia created a whole new world.

The expression "exponential growth" is often used to describe our age, but the trouble is that it has been used to describe just about every age since the invention of exponentials. In every age, there are always some things that grow exponentially, but others don't. For every technological innovation there was a moment when it spread "exponentially", whether it was church clocks or windmills, reading glasses or steam engines; and their "quality" improved exponentially for a while, until the industry matured or a new technology took over. Moore's law is nothing special: similar exponential laws can be found for many of the old inventions. Think how quickly radio receivers spread: in the USA there were only five radio stations in 1921 but already 525 in 1923. Cars? The USA produced 11,200 in 1903, but already 1.5 million in 1916. By 1917 a whopping 40% of households had a telephone in the USA up from 5% in 1900. There were fewer than one million subscribers to cable television in 1984, but more than 50 million by 1989. The Wright brothers flew the first airplane in 1903: during World War i (1915-18) France built 67,987 airplanes, Britain 58,144, Germany 48,537, Italy 20,000 and the USA 15,000, for a grand total of almost 200 thousand airplanes; after just 15 years of its invention. In 1876 there were only 3,000 telephones: 23 years later there were more than a million. Neil Armstrong stepped on the Moon in 1969, barely eight years after Yuri Gagarin had become the first human to leave the Earth's atmosphere.

Most of these fields then slowed down dramatically. And 47 years after the Moon landing we still haven't sent a human being to any planet and we haven't even returned to the Moon since the Apollo 17 in 1972. Similar statistics of "exponential growth" can be found for other old inventions, all the way back to the invention of writing. Perhaps each of those ages thought that growth in those fields would continue at the same pace forever. The wisest, though, must have foreseen that eventually growth starts declining in every field. Energy production increased 13-fold in the 20th century and freshwater consumption increased 9-fold, but today there are many more experts worried about a decline (relative to demand) than experts who believe in one more century of similar growth rates.

Furthermore, there should be a difference between "change" and "progress". Change for the sake of change is not necessarily "progress". Most "updates" in my software applications have negative, not positive effects, and we all know what it means when our bank announces "changes" in policies. If i randomly change all the cells in your body, i may boast of "very rapid and dramatic change" but not necessarily of "very rapid progress". Assuming that any change equates with progress is not only optimism: it is the recipe for ending up with exactly the opposite of progress. Out of the virtually infinite set of possible changes, only a tiny minority, a tiny subset, would constitute progress.

There has certainly been progress in telecommunications; but what difference does it make for ordinary people whether a message is sent in a split second or in two split seconds? In 1775 it took 40 days for the English public to learn that a revolution had started in the American colonies. Seven decades later, thanks to the telegraph, it took minutes for the news of the Mexican War to travel to Washington. That is real progress: from 40 days to a few minutes. The telegraph did indeed represent "exponential" progress. Email, texting and chatting have revolutionized the way people communicate over long distances, but it is debatable whether that is (quantitatively and qualitatively) the same kind of revolution that the telegraph and the telephone caused.

There are many "simpler" fields in which we never accomplished what we set out to accomplish originally, and pretty much abandoned the fight after the initial enthusiasm. We simply became used to the failure and forgot our initial enthusiasm. For example, domestic lighting progressed dramatically from gas lighting to Edison's light bulbs and Brush's arc lights of the 1880s and the first tungsten light-bulbs and then to the light-bulbs of the 1930s, but since then there has been very little progress: as everybody whose eyesight is aging knows too well, we still don't have artificial lighting that compares with natural sunlight, and so we need to wear reading glasses in the evening to read the same book that we can easily read during the day. A century of scientific and technological progress has not given us artificial lighting that matches sunlight.

I can name many examples of "change" that are often equated with "progress" when in fact it is not clear what kind of progress it is bringing. The number of sexual partners that a person has over a lifetime has greatly increased, and social networking software allows one to have thousands of friends all over the world, but i am not sure that these changes (that qualify as "progress" from a strictly numerical point of view) result in happier lives. I am not sure that emails and text messages create the same bonds among people than the phone conversation, the letter on paper, the postcard and the neighbor's visit did.

One can actually argue that there is a lot of "regress", not "progress". We now listen to lo-fi music on computers and digital music players, as opposed to the expensive hi-fi stereos that were commonplace a generation ago. Mobile phone conversations are frequently of poor quality compared with the old land lines. We have access to all sorts of food 24 hours a day but the quality of that food is dubious. Not to mention "progress" in automated customer support, which increasingly means "search for the answer by yourself on the Web" (especially from high-tech software giants like Microsoft, Google and Facebook) as opposed to "call this number and an expert will assist you".

In the early days of the Internet (1980s) it was not easy to use the available tools but any piece of information on the Internet was written by very competent people. Basically, the Internet only contained reliable information written by experts. Today there might be a lot more data available, but the vast majority of what travels on the internet is: a) disinformation, b) advertising. It is not true that in the age of search engines it has become easier to search for information. Just the opposite: the huge amount of irrelevant and misleading data is making it more difficult to find the one webpage that has been written by the one great expert on the topic. In the old days her webpage was the only one that existed. (For a discussion on Wikipedia see the appendix).

Does the Internet itself represent true progress for human civilization if it causes the death of all the great magazines, newspapers, radio and television programs, the extinction of bookstores and record stores, and if it will make it much rarer and harder to read and listen to the voices of the great intellectuals of the era? while at the same time massively increasing the power of corporations (via targeted advertising) and of governments (via systemic surveillance)? From the Pew Research Center's "State of the News Media 2013" report: "Estimates for newspaper newsroom cutbacks in 2012 put the industry down 30% since its peak in 2000. On CNN, the cable channel that has branded itself around deep reporting, produced story packages were cut nearly in half from 2007 to 2012. Across the three cable channels, coverage of live events during the day, which often require a crew and correspondent, fell 30% from 2007 to 2012... Time magazine is the only major print news weekly left standing".

Even the idea that complexity is increasing relies on a weak definition of "complexity". The complexity of using the many features of a smartphone is a luxury and cannot be compared with the complexity of defending yourself from wild animals in the jungle or even with the complexity of dealing with weather, parasites and predators when growing food in a farm. The whole history of human civilization is a history of trying to reduce the complexity of the world. Civilization is about creating stable and simple lives in a stable and simple environment. By definition, what we call "progress" is a reduction in complexity, although to each generation it appears as an increase in complexity because of the new tools and the new rules that come with those tools. Overall, living has become simpler (not more complicated) than it was in the stone age. If you don't believe me, go and camp in the wilderness by yourself with no food and only stone tools.

In a sense, today's Singularity prophets assume that machine "intelligence" is the one field in which growth will never slow down, but will keep accelerating forever.

Again, i would argue that it is not so much "intelligence" that has accelerated in machines (their intelligence is the same that Alan Turing gave them when he invented his "universal machine") as much as miniaturization. Moore's law (which was indeed exponential while it lasted) had nothing to do with machine intelligence, but simply with how many transistors one can squeeze on a tiny integrated circuit. There is very little (in terms of intelligent tasks) that machines can do today that they could not have done in

1950 when Turing published his paper on machine intelligence. What has truly changed is that today we have extremely powerful computers squeezed into a palm-size smartphone at a fraction of the cost. That's miniaturization. Equating miniaturization to intelligence is like equating an improved wallet to wealth.

Which progress really matters for Artificial Intelligence: hardware or software? There has certainly been rapid progress in hardware technology (and in the science of materials in general) but the real question to me is whether there has been any real progress in software technology since the invention of binary logic and of programming languages. And a cunning software engineer would argue that even that question is not correct: there is a difference between software engineering (that simply finds ways to implement algorithms in programming languages) and algorithms. The computer is a machine that executes algorithms. Anybody trying to create an intelligent machine using a computer is trying to find the algorithm or set of algorithms that will match or surpass human intelligence. Therefore it is neither progress in hardware nor progress in software that really matters (those are simply enabling technologies); what matters is progress in Computational Mathematics.

Ray Kurzweil's book used a diagram titled "Exponential Growth in Computing", but i would argue that it is bogus because it starts with the electromechanical tabulators of a century ago: it is like comparing the power of a windmill to the power of a horse. Sure there is an exponential increase in power, but it doesn't mean that windmills will keep improving forever vis à vis horsepower and windpower. And it doesn't distinguish between progress in hardware and progress in software, nor between progress in software and progress in algorithms. What we would like to see is a diagram titled "Exponential Growth in Computational Math". As i am writing this, most A.I. practitioners are looking for abstract algorithms that improve automatic learning techniques.

Others believe that the correct way to achieve artificial intelligence should be to simulate the brain's structure and its neural processes, a strategy that greatly reduces the set of interesting algorithms. In that case, one would also want to see a diagram titled "Exponential Growth in Brain Simulation". Alas, any neurologist can tell you how far we are from understanding how the brain performs even the simplest daily tasks. Current brain simulation projects are modeling only a small fraction of the structure of the brain, and provide only a simplified binary facsimile of it: neuronal states are represented as binary states, the variety of neurotransmitters is reduced to just one kind, the emphasis is on feed-forward rather than on feedback connections, and, last but not least, there is usually no connection to a body. No laboratory has yet been able to duplicate the simplest brain we know, the brain of the 300-neuron roundworm: where's the exponential progress that would lead to a simulation of the 86 billion-neuron brain of Homo Sapiens (with its 100 trillion connections)? Since 1963 (when Sydney Brenner first proposed it), scientists worldwide have been trying to map the neural connections of the simplest roundworm, the Caenorhabditis Elegans, thus jump-starting a new discipline called Connectomics. So far they have been able to map only subsets of the worm's brain responsible for specific behaviors.

If you believe that an accurate simulation of brain processes will yield artificial intelligence (whatever your definition is of "artificial intelligence"), how accurate has that simulation to be? This is what neuroscientist Paul Nunez has called the "blueprint problem". Where does that simulation terminate? Does it terminate at the computational level, i.e. at simulating the exchanges of information within the brain? Does it terminate at the molecular level, i.e. simulating the neurotransmitters and the very flesh of the brain? Does it terminate at the electrochemical level, i.e. simulating electromagnetic equations and chemical reactions? Does it terminate at the quantum level, i.e. taking into consideration subatomic effects?

Ray Kurzweil's "Law of Accelerating Returns" is nothing but the usual enthusiastic projection of the present into the future, a mistake made by millions of people all the time. Alas, millions of people buy homes when home values are going up believing that they would go up forever. Historically, most technologies grew quickly for a while, then stabilized and continued to grow at a much slower pace until they became obsolete.

We may even overestimate the role of technology. Some increase in productivity is certainly due to technology, but in my opinion other contributions have been neglected too quickly. For example, Luis Bettencourt and Geoffrey West of the Santa Fe Institute have shown that doubling the population of a city causes on average an increase of 130% in its productivity ("A Unified Theory of Urban Living", 2010). This has nothing to do with technological progress but simply with urbanization. The rapid increase in productivity of the last 50 years may have more to do with the rapid urbanization of the world than with Moore's law: in 1950 only 28.8% of the world's population lived in urban areas but in 2008 for the first time in history more than half of the world's population lived in cities (82% in North America, the most urbanized region in the world).

Predictions about future exponential trends have almost always been wrong. Remember the prediction that the world's population would "grow exponentially"? In 1960 Heinz von Foerster predicted that population growth would become infinite by Friday the 13th of November 2026. Now we are beginning to fear that it will actually start shrinking (it already is in Japan and Italy). Or the prediction that energy consumption in the West will grow exponentially? It peaked a decade ago; and, as a percentage of GDP, it is actually declining rapidly. Life expectancy? It rose rapidly in the West between 1900 and 1980 but since then it has barely moved. War casualties were supposed to grow exponentially with the invention of nuclear weapons: since the invention of nuclear weapons the world has experienced the lowest number of casualties ever (see Steven Pinker's book "The Better Angels of Our Nature"), and places like Western Europe, that had been at war nonstop for 1500 years, have not had a major war since 1945.

There is one field in which i have witnessed rapid (if not exponential) progress: Genetics. This discipline has come a long way in just 70 years, since Oswald Avery and others identified DNA as the genetic material (1944) and James Watson and Francis Crick discovered the double-helix structure of DNA (1953). Frederick Sanger produced the first full genome of a living being in 1977, Kary Banks Mullis developed the polymerase chain reaction in 1983, Applied Biosystems introduced the first fully automated sequencing machine in 1987, William French Anderson performed the first procedure of gene therapy in 1990, Ian Wilmut cloned a sheep in 1997, the sequencing of the human genome was achieved by 2003, and Craig Venter and Hamilton Smith reprogrammed a bacterium's DNA in 2010. The reason that there has been such dramatic progress in this field is that a genuine breakthrough happened with the discovery of the structure of DNA. I don't believe that there has been an equivalent discovery in the field of Artificial Intelligence.

Economists would love to hear that progress is accelerating because it has an impact on productivity, which is one of the two factors driving GDP growth. GDP growth is basically due to population growth plus productivity increase. Population growth is coming to a standstill in all developing countries (and declining even in countries like Iran and Bangladesh) and, anyway, in the 20th century the biggest contributor to workforce growth was actually women, which came to the workplace by the millions, but now that number has stabilized.

If progress were accelerating, you'd expect productivity growth to accelerate. Instead, despite all the hoopla about computers and the Internet, productivity growth of the last 30 years has averaged 1.3% compared to 1.8% in the previous 40 years. Economists like Jeremy Grantham now predict a future of zero growth ("On The Road To Zero Growth," 2012). Not just deceleration but a shrieking halt.

Whenever i meet someone who strongly believes that machine intelligence is accelerating under our nose, i ask him/her a simple question: "What can machines do today that they could not do five years ago?" If their skills are "accelerating" and within 20-30 years they will have surpassed human intelligence, it shouldn't be difficult to answer that question. So far the answers to that question have consistently been about incremental refinements (e.g., the new release of a popular smartphone that can take pictures at higher resolution) and/or factually false ("they can recognize cats", which is not true because in the majority of cases these apps still fail, despite the results of the ImageNet Competitions).

In 1939 at the World's Fair in New York the General Motors Futurama exhibit showed how life would be in 1960 thanks to technological progress: the landscape was full of driverless cars. The voiceover said: "Does it seem strange? Unbelievable? Remember, this is the world of 1960!" Twentyone years later the world of 1960 turned out to be much more similar to the world of 1939 than to the futuristic world of that exhibit.

On the 3rd of April 1988 the Los Angeles Times Magazine ran a piece titled "L.A. 2013" in which experts predicted how life would look like in 2013. They were comfortable predicting that the average middle-class family would have two robots to carry out all household chores including cooking and washing; that kitchen appliances would be capable of intelligent tasks; and that people would commute to work in self-driving cars. How many robots do you have in your home and how often do you travel in a self-driving car?

In 1964 Isaac Asimov wrote an article in the New York Times (August 16) titled "Visit to the World's Fair of 2014" in which he predicted what the Earth would look like in 2014. He envisioned that by 2014 there would be Moon colonies and all appliances would be cordless.

I am told that you must mention at least one Hollywood movie in a book on A.I. The only one that deserves to be mentioned is "2001: A Space Odyssey" (1968) by Stanley Kubrick. It is based on a book by Arthur Clarke. It features the most famous artificial intelligence of all times, HAL 9000. In the book HAL was born in 1997. 1997 came and went with no machines even remotely capable of what HAL does in that film.

The future is mostly disappointing. As Benjamin Bratton wrote in December 2013: "Little of the future promised in TED talks actually happens".

People who think that progress has been dramatic are just not aware of how fast progress was happening before they were born and of how high the expectations were and of how badly those expectations have been missed by current technology. Otherwise they would be more cautious about predicting future progress.

Intermezzo: In Defense of Regress

We accept as "progress" many innovations whose usefulness is dubious at best. Here are some favorite examples.

Any computer with a "mouse" requires the user to basically have three hands.

Never since the 1950s have phone communications been so rudimentary as after the introduction of the mobile phone. Conversations invariably contain a lot of "Can you hear me?" like in the age of black and white movies. I felt relieved in a Mexican town where there was a public phone at every corner: drop a coin and you are making a phone call. Wow. No contract needed, and no "can you hear me?"

Mobile phone ringers that go off in public places such as movie theaters and auditoria (and that obnoxiously repeat the same music-box refrain in a mechanical tone) do not improve the experience.

Voice recognition may have represented an improvement when it allowed people to say numbers rather than press them on the phone keyboard; but now the automated system on the other side of the phone asks you for names of cities or even "your mother's maiden name", and never gets them right (especially if you, like me, have a foreign accent) or for long numbers (such as the 16-digit number of my credit card) that you have to repeat over and over again until it gets it right or it gives up and mercifully connects you to a human operator.

The automation of cash registers means that it takes longer to pay than to find the item you want to buy (and you cannot buy it at all if the cash register doesn't work).

The car keys with an embedded microchip (the "transponder" keys) cost 140 times more to duplicate than the old chip-less car keys.

Watching films on digital media such as DVDs is more difficult for a professional critic than watching them on videotapes because stopping, rewinding, forwarding and, in general, pinpointing a scene is much easier and faster on analog videotapes (VCRs) than on digital files.

Computer's and car's CD drives that you have to push (instead of pull) in order to open are simply more likely to break and don't really add any useful feature. If the CD or DVD gets stuck inside, the drive can only be opened with a special screwdriver that virtually no user has.

Most portable gadgets used to operate with the same AA or AAA batteries. When on the road, you only had to worry about having those spare batteries. Now most cameras work only with proprietary rechargeable batteries: the fact that they are "rechargeable" is useless if they die in a place where you cannot recharge them, which is the case whenever you are far from a town or forgot the charger at home. I don't see this as progress compared with the cheap, easily replaceable AA batteries that i could also use with my hiking GPS, my headlight and my walkie-talkie. In fact, Nikon mentions it as a plus that its Coolpix series is still "powered by readily available AA batteries".

It is hard to believe that there was a time (a century ago) when you would pick up the phone and ask an operator to connect you to someone. Now you have to dial a 10-digit number, and sometimes a 13-digit number if you are calling abroad. More recently there used to be telephone directories to find the phone number of other telephone subscribers. I remember making fun of Moscow when we visited it in the 1980s because it didn't have a telephone directory. In the age of mobile phones the telephone directory has disappeared: you can know a subscriber's number only if someone gives it to you. Apparently the Soviet Union was the future, not the past.

Thanks to air-conditioned buildings with windows that are tightly sealed, we freeze in the summer and sometimes catch bronchitis while it is really hot outside.

Talking of windows, the electric windows of your car won't operate if the car's battery dies (the old "roll down the window" does not apply to a car with dead battery).

In most of the developed world, when you travel by bus or train, you need to get your ticket at a machine or have exact change to buy it on the bus, hardly an improvement over the old system of paying the conductor when you board. New buses and trains are climatized: it is impossible to take decent pictures of the landscape because the windows cannot be opened and are dimmed.

Printing photographs has become more, not less, expensive with the advent of digital cameras, and the quality of the print is debatable.

The taximeter, rarely used in developing countries but mandatory in "advanced" countries, is a mixed blessing. Basically, a taxi driver asks you to buy a good without telling you the price until you have already used the good and you cannot change your mind. The taximeter often increases the cost of a ride because you can't bargain anymore as you would normally do based on the law of supply and demand (for example, in situations when the taxi driver has no hopes for other customers). Furthermore, the taximeter motivates unscrupulous taxi drivers to take the longest and slowest route to your destination, whereas a negotiated price would motivate the driver to deliver you as quickly as possible.

Thanks to "progress" in software, over the years i had to adapt to countless limitations. Since 1983 the name on my email address has been this:

piero .•´¯`•.¸.•´¯`•..•´¯`•. scaruffi

In 2016 i was informed that a new release of the email system does not admit special characters.

After being notified a thousand times by a very aggressive Windows 10 operating system that new updates were available, one day i finally clicked on Yes and... Movie Maker stopped working: it now consistently objects that my brand new laptop does not meet the minimum requirements (yes, it does). A few days later i received another notification that new updates were available and immediately clicked on Yes hoping that one of these updates would fix the problem that keeps Movie Maker from running. The only noticeable difference is that now my laptop arranges all the icons to the left, no matter how i try to arrange them. I tried to get rid of the annoying "lock screen". I searched the Web and found that thousands of Windows 10 users are as annoyed as me by this "feature". There is absolutely no information on the Microsoft website but there are forums ("customer support" in the age of intelligent machines) where several people have posted a solution that worked for me. Quote:

- Open the registry editor.
- Navigate to
HKEY_LOCAL_MACHINE\SOFTWARE\Policies\Microsoft\Windows
- Create a new registry key called Personalization
- Navigate to the Personalization key
- Right click in the right pane and select New then DWORD (32-bit) Value.
- Name the new value NoLockScreen
- Set NoLockScreen to 1"

This was titled "Simple Steps To Get Rid Of Windows 10's New Lock Screen".

No, this does not happen only with Windows 10, nor only with Microsoft software. It happens with all the software out there.

If progress means that what i have been using will not work anymore it is not progress. It is progress for the ones who make it and sell it, but not for the ones who never asked for it and are now forced to accept it and pay for it.

Computers can be amazingly unintuitive compared with older devices. If you remove a USB flash drive the way you normally remove a CD or DVD from its player, you may lose all the data, so you are required to "safely remove" it. On Apple computers the way to safely remove a drive is to… throw it in the garbage can!

Websites with graphics, animation, pop-up windows, "click here and there", cause you to spend most of the time scrolling away from these digital paraphernalia instead of reading the information that you were looking for.

We the consumers passively accept too many of these dubious "improvements".

Most of these "improvements" may represent progress, but the question is "progress for whom"? Pickpockets used to steal one wallet at a time. The fact that today a hacker can steal millions of credit card numbers in an instant constitutes progress, but progress for whom?

And don't get me started on "health care", which in these high-tech days has become less about "health" and more and more about making you chronically ill. You are perfectly fine until you walk into the office of a dentist, eye doctor or other specialist; when you come out, you have become a medication addict with an immune system weakened by antibiotics and some prosthetic addition to your body that will require lifelong maintenance: progress for the "health-care" industry, not for you. (In 2016 a study published in the British Medical Journal by Martin Makary, a surgeon at the Johns Hopkins University, estimated that medical error was the third leading cause of death in the USA after heart disease and cancer).

We live in a world of not particularly intelligent machines, but certainly in a world of machines that like to beep. My car beeps when I start it, when I leave the door open, and if I don't fasten my seat belt. Note that it doesn't beep if something much more serious happen, like the alternator dies or the oil level is dangerously low. My microwave oven beeps when the food is ready, and it keeps beeping virtually forever unless someone opens its door (it doesn't matter that you actually pick up the food, just open the door). My printer beeps when it starts, when it needs paper, and whenever something goes wrong (a blinking message on the display is not enough, apparently). Best of all, my phone beeps when i turn it off, and, of course, sometimes i turn it off because i want it silent: it will beep to tell everybody that it is being silenced. I think that every manual should come with instructions on how to disable the beeping on the device: "First and foremost, here is how you can completely shut up your device once and forever".

Last but not least, something is being lost in the digital age, something that was the fundamental experience of (broadly defined) entertainment. During a vacation in a developing country i watched as a girl came out of the photographer's shop. She couldn't wait and immediately opened the envelope that contained her photographs. I witnessed her joy as she flipped through the pictures. The magic of that moment, when she sees how the pictures came, will be gone the day she buys her first digital camera. There will be nothing special about watching the pictures on her computer's screen. There will be no anxious waiting while she uploads them to the computer because, most likely, she will already know how the pictures look like before she uploads them. Part of the magic of taking photographs is gone forever, replaced by a new, cold experience that consists in refining the photograph with digital tools until it is what you want to see, not what it really looked like, and then posting it on social media in an act of vanity.

Or take live events. The magic of a live sport event used to be the anxious wait for the event to start, and then the "rooting" for one of the competitors or teams. After the introduction of TiVo, one can watch a "live" event at any time by conveniently "taping" it. Many live events are actually broadcasted with a slight delay, so you may find on the Internet the result of a soccer game that is still going on according to your television channel. Thus the "waiting" and the "rooting" are no longer the two fundamental components of the "live" experience. The whole point of watching a live event was the irrational feeling that your emotional state could somehow influence the result. If the event is recorded (i.e., is already in the past), that feeling disappears and you have to face the crude reality of your impotence to affect the result. But then what's the point of rooting? Thus the viewer is unlikely to feel the same emotional attachment to the game that s/he is watching. In the back of her/his mind, it is clear that the game has already finished. The experience of watching the "live" event is no longer one of anxiety but one of appreciation. Told by a friend that it was a lousy game, the viewer may well decide not to watch the event that her home appliance taped.

Yes, i know that Skype and Uber and many new services can solve or will solve these problems, but the point is that these gadgets and features were conceived and understood as "progress" when they were introduced (and usually amid much fanfare). The very fact that platforms such as Skype and Uber have been successful proves that the quality of services in those fields had overall regressed, not progressed, and therefore there was an opportunity for someone to restore service to a decent level.

We should always pause and analyze whether something presented as "progress" truly represents progress. And for whom.

Intermezzo: Why Futurists Always Get it Wrong

Because they want to predict the future without first studying the past.

And because they underestimate how important society is to shape the future, as opposed to "exponential" technological progress. It is the eccentric in the garage, thinking of something completely different from the mainstream, who writes the history of the future. No futurist predicted Gutenberg, Columbus, Watt, Mendel, Edison, Marconi, Einstein, Fleming, Turing, Crick, Berners-Lee, Wozniak, Page, Zuckerberg... the scientists and inventors who truly changed the world.

In 1963 John McCarthy founded the Stanford AI Lab (SAIL) with the goal of building a fully intelligent machine within a decade. In 1965 Herbert Simon predicted that "machines will be capable, within twenty years, of doing any work a man can do". In 1967

Marvin Minsky predicted that "within a generation... the problem of creating artificial intelligence will substantially be solved", and anticipated that solving the problem of computer vision would take only a summer. In 1978 Moravec predicted that computers would become as intelligent as human beings in 1998 (but then in 1998 he published the essay "When Will Computer Hardware Match the Human Brain?"). I am still to find a single prediction by Kurzweil that turned out to be true, certainly none of those listed in "The Age of Spiritual Machines" (1999), except for those that everybody was already predicting (but this didn't stop an anonymous Wikipedia article from crediting him with a success rate of 86%). One of my favorites is Gartner Group's 2007 prediction that a whopping 80% of Internet users would participate in virtual worlds by 2012. The year 2012 came and went, and to this day (2016) the vast majority of Internet users do not even know what a virtual world is.

According to Stewart Brand, Marvin Minsky believed that contemporary philosophers were "shallow and wrong", but of course that could have been because contemporary philosophers proved him (Minsky) shallow and wrong.

Jobs in the Age of the Robot – Part 1: What Destroys Jobs

During the Great Recession that ravaged the Western world in 2008-2011, both analysts and ordinary families were looking for culprits to blame for the high rate of unemployment, and automation became a popular one in the developed world. Automation was indeed responsible for making many jobs obsolete, but it was not the only culprit nor the main one.

The first and major factor that accounts for the demise of many jobs in the Western world is the end of the Cold War. Before 1991 the economies that really mattered were a handful (USA, Japan, Western Europe). Since 1991 the number of competitors for the industrialized countries has skyrocketed, and they are becoming better and better at competing with the West. Technology might have "stolen" some jobs, but that factor pales by comparison with the millions of jobs that were exported to Asia. In fact, if one considers the totality of the world, an incredible number of jobs have been created precisely during the period in which critics argue that millions of jobs have been lost to automation. If Kansas loses one thousand jobs but California creates two thousand, we consider it an increase in employment. These critics make the mistake of using the old nation-based logic for the globalized world. When counting jobs lost or created during the last twenty years, one needs to consider the entire interconnected economic system that spreads all over the planet. Talking about the employment data for the USA but saying nothing about the employment data (over the same period) of China, India, Mexico and so forth is distorting the picture. If General Motors lays off one thousand employees in Michigan but hires two thousand in China, it is not correct to simply conclude that "one thousand jobs have been lost". If the car industry in the USA loses ten thousand jobs but the car industry in China gains twenty thousand, it is not correct to simply conclude that ten thousand jobs have been lost by the car industry. In these cases jobs have actually been created.

That was precisely the case: millions of jobs were created by the USA in the rest of the world while millions were lost at home. The big driver was not automation but, cheap labor.

Then there are sociopolitical factors. Unemployment is high in Western Europe, especially among young people, not because of technology but because of rigid labor laws and government debt. A company that cannot lay off workers is reluctant to hire any. A government that is indebted cannot pump money into the economy. This is a widespread problem in the Western economies of our age. It has to do with politics, not with automation.

Germany is as technologically advanced as the USA. All sorts of jobs have been fully automated. And, still, in Germany the average hourly pay has risen five times faster between 1985 and 2012 than in the USA. This has little to do with automation: it has to do with the laws of the country. Hedrick Smith's "Who Stole the American

Dream?" (2012) lays the blame on many factors, but not on automation.

In 1953 Taiichi Ohno invented "lean manufacturing" at Japan's Toyota, possibly the most important revolution in manufacturing since Ford's assembly line. Nonetheless, Japan created millions of jobs in manufacturing; and, in fact, Toyota went on to become the largest employer in the world of car-manufacturing jobs. Even throughout its two "lost decades" (1991-2010) Japan continued to post very low unemployment. Japan has perhaps the highest number of industrial robots of any country, and it also enjoys one of the lowest unemployment rates in the world. Germany is a close second in automation, and it has the lowest unemployment figures for any major country in Western Europe.

Another major factor that accounts for massive losses of jobs in the developed world is the management science that emerged in the 1920s in the USA. That science is the main reason that today companies don't need as many employees as comparable companies employed a century ago. Each generation of companies has been "slimmer" than the previous generation. As those management techniques get codified and applied across all departments, companies become more efficient at manufacturing (world-wide), at selling (using the most efficient channels) and at predicting business cycles. All of this results in fewer employees not because of automation but because of optimization.

In May 2016 Challenger, Gray & Christmas estimated the companies that had laid off the most workers. The top job cutter of the first four months of 2016 was National Oilwell Varco, a Texan company making equipment for the petroleum industry. Job cutters #3 (Schlumberger), #5 (Halliburton), #7 (Chevron) and #10 (Weatherford) were all involved in the petroleum business. This had nothing to do with robots or artificial intelligence but simply with record-low oil prices. Walmart was the second job-cutter in the country, but, like all retail chains, its problem was simply the competition from online sales. Meanwhile, the US economy was adding about 200,000 new jobs each month, and those jobs were consistently in high-tech sectors. Intel (#4) and Dell (#6) too were in that list. Both missed the mobile revolution and were being replaced by other firms. Their job cutting was not due to more automation in the factories.

Additionally, in the new century the USA has deliberately restricted immigration to the point that thousands of brains are sent back to their home countries even after they graduated in the USA. This is a number that is virtually impossible to estimate, but, in a free market like the USA that encourages innovation and startups, jobs are mostly created via innovation, and innovation comes from the best brains, which account for a tiny percentage of the population. Whenever the USA sends back or refuses to accept a foreign brain that may become one of those creators of innovation, the USA is de facto erasing thousands of future jobs. Those brains are trapped in places where the system does not encourage the startup-kind of innovation or where capital is not as readily available. They are wasted in a way that equivalent brains were not wasted in the days when immigration into the USA was much easier, up until the generation of Yahoo, eBay and Google. According to a study contained in the "Kauffman Thoughtbook 2009" by the Kauffman Foundation, foreign-born entrepreneurs ran 24% of the technology businesses started between 1980 and 1998 (in Silicon Valley a staggering 52%). In 2005 these companies generated $52 billion in revenue and employed 450,000 workers. In 2011 a report from the Partnership for a New American Economy found that 18% of the Fortune 500 companies of 2010 were founded by immigrants. These companies had combined revenues of $1.7 trillion and employed millions of workers. If one includes the Fortune 500 companies founded by children of immigrants, the combined revenues were $4.2 trillion in 2010, greater than the GDP of any other country in the world except China and Japan.

Technology is certainly a factor, but it can go either way. Take, for example, energy. This is the age of energy. Energy has always been important for economic activity but never like in this century. The cost and availability of energy are major factors to determine growth rates and therefore employment. The higher the cost of energy, the lower the amount of goods that can be produced, the lower the number of people that we employ. If forecasts by international agencies are correct, the coming energy boom in the USA (see the International Energy Agency's "World Energy Outlook" of 2012) will create millions of jobs, both directly and indirectly. That energy boom is due to new technology.

When the digital communication and automation technologies first became widespread, it was widely forecast a) that people would start working from home and b) that people would not need to work as much. What i have witnessed is the exact opposite: virtually every company in Silicon Valley requires people to show up at work a lot more than they did in the 1980s, and today virtually everybody is "plugged in" all the time. I have friends who check their email and text messages all the time while we are driving to the mountains and even while we are hiking. The digital communication and automation technologies have not resulted in machines replacing these engineers but in these engineers being able to work all the time from everywhere, and sometimes their companies require it. Those technologies have resulted in people working a lot more. (The willingness of people to work more hours for free is another rarely mentioned factor that is contributing to higher unemployment).

Jobs in the Age of the Robot – Part 2: What Creates Jobs

Unemployment cannot be explained simply by looking at the effects of technology. Technology is one of many factors and, so far, not the main one. There have been periods of rapid technological progress that have actually resulted in very low unemployment (i.e. lots of jobs), most recently in the 1990s when e-commerce was introduced, despite the fact that the digital camera had killed the photographer's shop, Amazon the bookstore, the mobile phone the land lines and Craigslist the local newspaper.

The effect of a new technology on employment is not always obvious, and that's why our first reaction is of fear. For example, who would have imagined that the technology of computers (invented for fast computation) would have created millions of new jobs in the sector of telecommunications?

A 2014 report by the Kauffman Foundation showed that between 1988 and 2011 almost all of new jobs were created by businesses less than five years old, while existing firms were net job destroyers, losing 1 million jobs net combined per year. By contrast, in their first year, new firms add an average of 3 million jobs."

New technologies also create jobs in other sectors. It is called the "multiplier effect". The people employed in the new technology need shops, restaurants, doctors, lawyers, schoolteachers, etc. A 2016 report by the Bay Area Council Economic Institute shows that the biggest multiplier effect of our times came from the high-tech industry: for each job created in the high-tech sector, more than four jobs are created in other sectors. A company that hires a software engineer is indirectly creating 4 new jobs in the community. By comparison, traditional manufacturing has a multiplier effect of 1.4 jobs.

Historically, in fact, technology created jobs while simultaneously destroying old jobs, and the new jobs have typically been better-paying and safer than the old ones. Not many people dream of returning to the old days when agriculture was fully manual and millions of people were working in terrible conditions in the fields. Today a few machines can water, seed, rototill and reap a large field. Those jobs don't exist anymore, but many jobs have been created in manufacturing sectors for designing and building those machines. In the USA of the 19th century, 80% of jobs were in agriculture; today only about 2% are. Yet it is not true that the mechanization of agriculture has caused 78% of people to remain unemployed. Few peasants in the world would like their children to grow up to be peasants instead of mechanical engineers. Ditto for computers that replaced typewriters and typewriters that replaced pens and pens that replaced human memory. Gutenberg's printing press put a few thousand scribes out of business, but it generated a mass production of books, which, besides educating the public and creating an infinite number of new businesses for educated people (e.g. magazines and newspapers), created millions of jobs to print books, market them and sell them. For each scribe that went out of business, thousands of bookstores popped up all over the world. Steam engines certainly hurt the horse and mule business, but created millions of jobs in factories and thousands of new businesses for the goods that could be made in those factories.

All of these forms of automation had side effects that were negative, but one negative side-effect that they did NOT have was to cause unemployment. They created more jobs than they destroyed, and better ones.

In the 1980s i worked in Silicon Valley as a software engineer and back then the general consensus was that software engineering was being automated and simplified at such a pace that soon it would become a low-paid job and mostly exported to low-wage countries like India. Millions of software jobs have in fact been "offsourced" to India, but the number of software developers in the USA has skyrocketed to 1,114,000 with a growth rate of 17% and an average salary of $100,000, which is more than twice the average salary of $ 43,643 (source: US Bureau of Labor Statistics, 2015).

It is true that the largest companies of the 21st century are much smaller than the largest companies of the 20th century. However, the world's 4,000 largest companies spend more than 50% of their revenues on their suppliers and a much smaller percentage on their people (as little as 12% according to some studies). Apple may not be as big as IBM was when it was comparable in power, but Apple is the reason that hundreds of thousands of people have jobs in companies that make the parts used in Apple's products.

I would be much more worried about the "gift economy": the fact that millions of people are so eager to contribute content and services for free on the Internet. For example, the reason that journalists are losing their jobs has little to do with the automation in their departments and a lot to do with the millions of amateur "bloggers" who provide content for free on the Internet.

If we take into account the global effects of automation, we reach different conclusions about the impact that robots (automation in general) will have. In the USA robots are likely to bring back jobs. The whole point of exporting jobs to Asia was to benefit from the lower wages of Asian countries; but a robot that works for free 24 hours a day 7 days a week beats even the exploited workers of communist China. As they become more affordable, these "robots" (automation in general) will displace Asian workers, not Michigan workers. The short-term impact will be to make outsourcing of manufacturing an obsolete concept. The large corporations that shifted thousands of jobs to Asia will bring them back to the USA. In the mid-term this could even have the secondary effect of putting Asian products out of the market and of creating a manufacturing boom in the USA: not only old jobs will come back but a lot of new jobs will be created. In the long term robots might create new kinds of jobs that today we cannot even foresee.

Let's take a simple example, a kind of robot that will appear soon at the supermarket of your neighborhood. As you enter the store, you will be welcomed by a mobile robot equipped with a basket and a check-out system. The robot will ask you what you are looking for, and escort you to the correct aisle of the store. It will let you browse the shelf and pick the brand you prefer. Then it will ask you to drop it in the basket. This will continue for as long as you have items to purchase. In fact, you could even read your shopping list to the robot and the robot will calculate and work out your passage through the store, optimally. For any product that you don't find the robot will investigate on the spot whether it can be ordered for you and delivered at your home address, or whether there is an affiliated store nearby where you can find it. The robot will also alert you to the products that are on sale and politely ask you if you'd like to take a look at them. When you are done, you will simply ask the robot "How much?" The robot will already know the total because it will have scanned each item that you dropped in its basket. The robot will take your payment (whether a credit card or a smartphone app) and print a receipt while escorting you back to your car. A robot like this is perfectly feasible today, except that it would still cost too much to build and operate, a cost not justified in stores that have a relatively low margin of profit on the goods that they sell. Nonetheless, who should panic at the prospect that such robots will someday exist? Which jobs are at risk? There is no human being who performs this task today. When you enter a store, you are on your own. If you have a question, good luck finding any employee who can help you. In many places the check-out operation is already a self-checkout. Hence, not a single job will be lost to these robots. On the other hand, imagine how many jobs will be created. Some company will become the Apple of shopping robots, and hire thousands of people to design, manufacture, sell and maintain these robots. Another company, the Google of robots, will come up with the idea of making home robots for shopping, robots that you keep in the house and drive with you to the store, and know the organization of each store connected to the cloud according to some Android-like standard. While you are walking into the store with your robot, your robot downloads the configuration of the store and the entire database of products, and then it starts behaving exactly as if it were that store's shopping robot. More jobs created here. Another company, the Oracle of robots, will come up with software that informs your robot about the best place for each item on a given shopping list. You won't even know where you are going. The robot will take you to a selection of stores that have what you need at the best prices. More jobs created. Then the Tesla of robots will come up with a way that you can 3D-print your custom shopping robot, as big or small as you want it, as environmental as you want it, as fast or slow as you want it. None of these jobs exist today. And almost no existing job is killed in this simple example.

Not many people in 1946 realized that millions of software engineers would be required by the computer industry in 2013. Robotics will require millions of robotics engineers. These engineers will not be as "smart" as their robots at whatever task for which those robots were designed just like today's software engineers are not as fast as the programs they create. As i type, Silicon Valley is paying astronomical salaries to robotics engineers and China is hiring thousands of people for the Internet of Things.

At the end of 2015 both the McKinsey report "Four fundamentals of workplace automation" and the study by James Bessen of Boston University's School of Law "How Computer Automation Affects Occupations" (November 2015) showed that robots will steal your job but will also create another job, and most likely it will be a better one in terms of income, health and personal satisfaction.

Bessen mentions the case of the ATM, one of the most successful programs to replace humans with machines. The routine jobs of bank tellers were very easy to automate, and, sure enough, the percentage of tellers in the USA fell from 20 per branch in 1988 to 13 in 2004. However, the banks reinvested the money that they saved and, in particular, they opened many more branches, that in turn hired many more tellers. ATMs ended up creating more jobs for tellers, and a huge number of jobs for companies building and maintaining ATMs, jobs that did not exist before.

Unfortunately, you do sell a lot of copies if you write books about the apocalypse, so writers are under pressure to be as negative as possible. Hence, bestsellers such as Martin Ford's "Rise of the Robots - Technology and the Threat of a Jobless Future" (2015) or Jerry Kaplan's "Humans Need Not Apply" (2015), which i thought were misleading, superficial and not helpful at all (to be fair, they both came out before those two reports).

According to the International Federation of Robotics, in 2016 the countries with the highest density of robot population were South Korea (478 robots per 10,000 workers), Japan (315) and Germany (292). These countries also had some of the lowest unemployment rates in the world. For the record, the number for the USA was 164 (the USA has higher, not lower, unemployment than South Korea and Japan), and the number for the euro-countries afflicted by chronic unemployment (such as Greece) was the lowest in the developing world. Data about Italy can be misleading: Italy has a relatively high density of robots but also high unemployment. However, almost all the robots are in the industrialized north, where unemployment is very low, and almost none in the south, where unemployment is one of the highest in the world. The number of robots sold in the United States increased by 43% in 2011 and has continued to increase rapidly, and unemployment has declined every single year since 2011. How in heaven did those distinguished writers conclude from the given data that robots cause unemployment?

It is always easy to imagine which jobs will be destroyed and very difficult to imagine the new jobs that technology will create. So we exaggerate the reality of the disappearing jobs and underestimate the reality of the new ones.

In 1992, one year after the invention of the first Internet browser, when newly elected president Bill Clinton assembled a group of experts to discuss the future of the economy, nobody mentioned the Internet (David Leonhardt, "The Depression - If Only Things Were That Good", New York Times, 2011).

The society of robots will create new jobs that today we can't even imagine. Robots will create an even more complex society in which human intelligence will be even more important. The future always surprises us.

And my guess is that robots will become obsolete too at some point, replaced by something else that today doesn't even have a name. Some day robots will be made obsolete by a new human invention. Robots will become obsolete way before humans become obsolete.

Jobs in the Age of the Robot – Part 3: The Sharing Economy

The real revolution in employment is coming from a different direction: the "sharing economy". Companies such as Airbnb, that matches people who own rooms and people looking for rooms to rent, and Uber, that matches drivers who own a car and people looking for a ride in a car, have introduced a revolutionary paradigm in the job market: let people monetize under-utilized assets. This concept will soon be applied in dozens of different fields, allowing ordinary people to find ordinary customers for their ordinary assets; or, in other words, to supply labor and skills on demand. Before the industrial revolution most jobs were in the countryside but urban industry existed and it consisted mainly of artisan shops. The artisans would occasionally travel to a regional market, but mostly it was the customer who looked for the artisan, not vice versa. Cities like Firenze (Florence) had streets devoted to specific crafts, so that a customer could easily find where all the artisans offering a certain product were located. Then came the age of the factory and of transportation, and industrialization created the "firm" employing thousands of workers organized in some kind of hierarchy. Having a job came to mean something else: being employed. Eventually society started counting "unemployed" people, i.e. people who would like to work for an employer but no employer wants their time or skills. The smartphone and the Internet are enabling a return of sorts to the model of the artisan era. Anybody can offer their time and skills to anybody who wants them. The "firm" is simply the intermediary that allows customers to find the modern equivalent of the artisan.

In a sense, the "firm" (such as Uber or Airbnb) plays the role that the artisan street used to play in Firenze. Everybody who has time and/or skills to offer can now become a "self-employed" person. And that "self-employed" person can work when she wants, not necessarily from 8 to 5. There is no need for an office and for hiring contracts.

The traditional firm has a workforce that needs to be fully employed all the time, and sometimes the firm has to lay off workers and sometimes has to hire some, according to complicated strategic calculations.

In the sharing economy, no such thing exists: the firm is replaced by a community of skilled workers who take the jobs they want to take, when they want to take them, and if the customer wants them to take them. In a sense, people can be fired and hired on the fly.

Of course, this means that "good jobs" will no longer be judged based on job promotions, salary increases and benefits. They will be based on customer demand (which in theory is what drives company's revenues which in turn drives job promotions, salary increases and benefits).

The unemployed person who finds it difficult to find a job in a firm is someone whose skill is not desired by any firm, but this does not mean that those skills are not desired by any customer. The firm introduced a huge interface between customer and worker. When there is a need for your skill, you have to hope that a manager learns of your skills, usually represented by a resume that you submitted to the human resources department, and hope that the financial officer will approve the hiring. The simple match-making between a customer who wants a service and the skilled worker who can provide that service gets complicated by the nature of the firm with its hierarchical structure and its system of checks and balances (not to mention internal politics and managerial incompetence). It would obviously be easier to let the customer deal directly with the skilled worker who can offer the required service.

Until the 2000s the problem was that the customer had no easy way of accessing skilled workers other than through the "yellow pages", i.e. the firms. Internet-based sharing systems remove the layers of intermediaries except one (the match-making platform, which basically provides the economy of scale). In fact, these platforms turn the model upside down: instead of a worker looking for employment in a firm that is looking for customers, the new model views customers as looking for workers. Not only does this model bypass the slow and dumb firm, but it also allows you to monetize assets that you own and you never perceived as assets. A car is an asset. You use it to go to work and to go on vacation, but, when it is parked in the garage, it is an under-utilized asset.

Marketing used to be a scientific process to shovel a new product down the throat of reluctant consumers: it now becomes a simple algorithm allowing customers to pick their skilled workers, an algorithm that basically combines the technology of online dating (match making), of auctions (bidding) and of consumer rating (that basically replaces the traditional "performance appraisal" prescribed in the traditional firm).

Of course, the downside of this new economy is that the worker has none of the protections that she had in the old economy: no security that tomorrow she will make money, no corporate pension plan, etc; and she is in charge of training herself to keep herself competitive in her business. The responsibility for a worker's future was mostly offloaded to the firm. In the sharing economy that responsibility shifts entirely to the worker herself.

The new proletariat is self-employed, and, basically, each member of the proletariat is actually a micro-capitalist; the price to pay is that the worker will have to shoulder the same responsibilities that traditionally have fallen into the realm of firm management.

People who worry about robots are thinking about the traditional jobs in the factory and the office.

Futurists have a unique way to completely miss the scientific revolutions that really matter.

Jobs in the Age of the Robot – Part 4: The Maid Principle

Older workers are scared at the prospect that their specialty skill will soon be performed by a machine. Students are scared at the prospect that they may be studying to perform a job that will not exist when they graduate. Both concerns are legitimate. Jobs will be created, but they will not be the jobs that we have today. Being able to adapt to new jobs, jumping from one skill to a very different skill, will make the difference between success and failure.

It is virtually impossible to give proper advice about jobs that don't exist today. It is difficult to imagine what foundations to study and what path to follow in order to be ready for a job that doesn't exist today. But here are some rules of thumb.

The most obvious among them is: anybody whose job consists in behaving like a machine will be replaced by a machine. In highly structured societies like the USA (where one cannot get an omelette in a restaurant after 11am despite the fact that they have all the ingredients in the kitchen and even the most inept of chefs certainly knows how to cook an omelette), many jobs fall into this category. Even the people who write press releases for big corporations, even speech writers, and to some extent even engineers are asked to follow rules and regulations. The higher the portion of their job that is governed by rules, the higher the chance that they will soon be replaced by a machine. Those same jobs are less vulnerable in countries where the "human touch" still prevails over clockwork organization.

Those people who are good at communicating, empathizing, and the other things that we expect from fellow humans, will not be replaced by machines any time soon. A nurse who simply performs a routine task and shows little or no emotional attachment to her patients will be replaced by a robot, but a nurse who also provides comfort, company and empathy is much harder to replace. There is no robot coming in the near future that can have a real conversation with a sick person or an elderly person.

If you behave and think like a machine, you are already redundant. There are many in the USA who fall into this category, people who get upset if we ask them to do something slightly different from what they have been trained to do. If you are one of those people who don't like to do anything that requires "thinking", ask yourself "why does the world need me?" Machines can do a better and friendlier job than you, with no lunch breaks, no sleep, no weekend parties and no exotic vacations. If you are a cog in a highly structured environment, you should be surprised that someone is still willing to pay you a salary.

On the other hand, if you are the one who designs the structured environment in which machines can thrive, or the one who designs the machines for that environment, or even just the one who builds, repairs and/or sells them, then we desperately need you, and we rely on you to make sure that machines will create a better world.

I actually like the idea that automation will keep challenging us to be more creative, to find higher and higher meanings to our lives. If machines can build a better world, why does the world need us? We have to answer this question. We are actually more human when we struggle to find a higher meaning to our lives than when we simply work from 8am till 5pm mindlessly following a routine as if we were… robots.

Interdisciplinary thinking will be more useful than ever and today's machines make it easier than before to get an interdisciplinary education. If you are using the power of machines, like your smartphone, to let machines do the thinking for you, you are probably getting dumber, and that will not help you. If you are using the power of machines to learn a lot more things than your parents did, in a lot more fields, you are more likely to compete for the best jobs of the future.

A serious problem in the USA is its increasingly under-educated population, that will certainly have trouble adjusting to the new job opportunities. This is already happening in sectors like software and biotech, where the new highly-paid jobs often go to the much better educated Chinese immigrants than to the native US citizens who dropped out of school. We immigrants of Silicon Valley can't help noticing that most of the people serving us in shops and restaurants were born and raised right here in the Bay Area, and totally missed the high-tech revolution that was happening under their nose. In 1946 the USA had the #1 high school graduation rate in the world. Today (according to the OECD) it ranks 22nd among 27 industrialized nations. US students rank 25th in mathematics, 17th in science and 14th in reading. Only 46% of US students finish college.

But there are also low-level jobs that cannot be automated easily. They cannot be automated because they are so "human". A favorite example is the hotel maid. This is a very low-wage job but which robot can pick objects of a virtually infinite range of shapes and solidity and use common sense to understand what must be done with them? Try to explain to a robot what "garbage" means. You don't throw dirty underwear away (it belongs to somebody) but you do throw empty pizza boxes away. On the other hand, you don't throw it away if the guest has written "maid: please save this" on it. If the empty pizza box contains dirty tissue paper, it is meant to be thrown away. But if the paper inside the empty pizza is green with the face of a president, maybe you should think twice.

"The main lesson of 35 years of AI research is that the hard problems are easy and the easy problems are hard" (Steven Pinker).

Marketing and Fashion

Back to the topic of accelerating progress: what is truly accelerating at exponential speed is fashion. This is another point where many futurists and high-tech bloggers confuse a sociopolitical phenomenon with a technological phenomenon.

What we are actually witnessing in many fields is a regression in quality. This is largely due to the level of sophistication reached by marketing techniques. Marketing is a scary human invention: it often consists in erasing the memory of good things so that people will buy bad things. There would be no market for new films or books if everybody knew about the thousands of good films and books of the past: people would spend their entire life watching and reading the (far superior) classics instead of the new films and books, most of which are mediocre at best. In order to have people watch a new film or read a new book, the marketing strategists have to make sure that people will never know about old films and books. It is often ignorance that makes people think they just witnessed "progress" in any publicized event. Often we call "progress" the fact that a company is getting rich by selling poor quality products. The "progress" lies in the marketing, not in the goods. The acceleration of complexity is in reality an acceleration of low quality.

We may or may not live in the age of machines, but we certainly live in the age of marketing. If we did not invent anything, absolutely anything, there would still be frantic change. Today change is largely driven by marketing. The industry desperately needs consumers to go out and keep buying newer models of everything. We mostly buy things we don't need. The younger generation is always more likely to be duped by marketing and soon the older generations find themselves unable to communicate with young people unless they too buy the same things. Sure: many of them are convenient and soon come to be perceived as "necessities"; but the truth is that humans have lived well (sometimes better) for millennia without those "necessities". The idea that an mp3 file is better than a compact disc which is better than a vinyl record is just that: an idea, and mainly a marketing idea. The idea that a streamed movie is better than a DVD which is better than a VHS tape is just that: an idea, and mainly a marketing idea. We live in the age of consumerism, of rapid and continuous change in products, mostly unnecessary ones.

What is truly accelerating is the ability of marketing strategies to create the need for new products. Therefore, yes, our world is changing more rapidly than ever; not because we are surrounded by better machines but because we are surrounded by better snake-oil peddlers (and dumber consumers).

"The computer industry is the only industry that is more fashion-driven than women's fashion" (I am quoting Larry Ellison, founder and chairman of Oracle).

Sometimes we are confusing progress in management, manufacturing and marketing (that accounts for 90 percent of the "accelerating progress" that we experience) with progress in machine intelligence (that is still at the "Press 1 for English" level).

Technological progress is, in turn, largely driven by its ability to increase sales. Therefore it is not surprising that the big success stories of the World-wide Web (Yahoo, Google, Facebook, etc) are the ones that managed to turn web traffic into advertising revenues. We are turning search engines, social media and just about every website into the equivalent of the billboards that dot city streets and highways. It is advertising revenues, not the aim of creating intelligent machines, that is driving progress on the Internet. In a sense, Internet technology was initially driven by the military establishment, that wanted to protect the USA from a nuclear strike, then by a utopian community of scientists that wanted to share knowledge, then by corporations that wanted to profit from e-commerce, and now by managers of advertising campaigns who want to capture as large an audience as possible. Whether this helps accelerate progress and in which direction is, at best, not clear.

When Vance Packard wrote his pamphlet "The Hidden Persuaders" (1957) on the advertising industry (on how the media can create the illusory need for unnecessary goods), he had literally seen nothing yet.

"The best minds of my generation are thinking about how to make people click ads" (I am quoting former Facebook research scientist Jeff Hammerbacher in 2012).

And, to be fair, the best minds of his generation are not only used to make people click on ads but also to create ever more sophisticated programs of mass surveillance (as revealed in 2013 by National Security Agency analyst Edward Snowden).

Recap

Here is what i have told you so far. There are assumptions underlying the belief that super-intelligent machines are coming soon. The first one is that A.I. is making staggering progress and the second one is that progress is accelerating like never before. I showed you that both statements are wild exaggerations. There are still colossal gaps in the program of A.I. and very few creative ideas on how to fill those gaps. Brute-force A.I. is unlikely to succeed outside problems of pattern recognition and brute-force A.I. has relied too much on faster and faster processors. Now that Moore's Law is coming to an end, we will need more (ahem, ahem) intelligent ways to do A.I. than brute force. I am not denying that there is progress in the way machines can work for us, but I will demystify why they can work better than in the past (it is more about the environment that we structure for the machines than about their intelligence). This explains why most of the machines around us are pretty stupid and why i don't see any robots walking around the streets of Silicon Valley. Precisely because of the limitations of today's A.I., you don't fear that machines will steal your job... unless your job is so stupid that even a stupid machine can do it.

As for the "accelerating progress" of our age, in most cases it is neither "unique" nor "progress". One century ago the world was completely changed by a series of inventions that happened one after the other within a few years: the telephone, the radio, the car, the airplane, the record, Quantum Mechanics, Relativity, etc. Are you sure that today's progress is more dramatic than that one? When answering this question, let's keep in mind that change is not always progress. Change can go in both directions: forward or backward. Change is not necessarily in the direction of progress. There have been a lot of changes in Syria since 2011, but only ISIS would call it "progress".

More criticism of today's A.I. is coming in the next pages, some of it more philosophical than practical, and in particular we will discuss the concept of super-human intelligence. But first let's go back to the fundamental question: why A.I. at all? And this will also link to what i wrote at the very beginning: I am not afraid of A.I.; i am afraid that it will not come soon enough.

Why we need A.I. or The Robots are Coming– Part 2: The Near Future of A.I. or Don't be Afraid of the Machine

The media are promising a myriad applications of A.I. in all sectors of the economy. So far we have seen very little compared to what was promised. In 2016 Bloomberg estimated 2,600 startups working on A.I. technology, but IDC calculated that sales for all companies selling A.I. software barely totaled $1 billion in 2015. There is a lot of talk, but, so far, very few actual products that people are willing to pay for.

The number-one application of A.I. is and will remain... drum roll... making you buy things that you don't need. All major websites employ some simple form of A.I. to follow you, study you, understand you and then sell you something. Your private life is a business opportunity for them and A.I. helps them figure out how to monetize it. The founders of A.I. are probably turning in their graves.

And sometimes these "things" can even kill you (the case of Wei Zexi in 2016, who was induced by an ad posted on Baidu to buy the cancer treatment that killed him).

Mark Weiser famously wrote: "The most profound technologies are those that disappear. They weave themselves into the fabric of everyday life until they are indistinguishable from it" ("The Computer for the 21st Century", 1991). Unfortunately, it turned out to be a prophecy about the ubiquitous "intelligent" agents that make us buy things.

Perhaps the most sophisticated (or, at least, widely used) A.I. system since 2014 is Facebook's machine-learning system FBLearner Flow, designed by Hussein Mehanna's team, that runs on a cluster of thousands of machines. It is used in every part of Facebook for quickly training and deploying neural networks. Neural networks can be fine-tuned by playing with several parameters. Optimizing these parameters is not trivial. It requires a lot of "trial and error". But even just a 1% improvement in machine-learning accuracy can mean billions of dollars of additional revenues for Facebook. So Facebook is now developing Asimo, that performs thousands of tests to find the best parameters for each neural network. In other words, Asimo does the job that is normally done by the engineers who build the deep-learning system.

While Jeff Hammerbacher's lament remains true, we must recognize that progress in deep learning has been driven by funding from companies like Google and Facebook whose main business interest is to convince people to buy things. If the world banned advertising from the Web, the discipline of deep learning would probably return to the obscure laboratories of the universities where it came from.

Remember Marshall McLuhan's comment in "Understanding Media" (1964) that "Far more thought and care go into the composition of any prominent ad in a newspaper or magazine than go into the writing of their features and editorials"? The same can be said today: far more thought and care has been invested in designing algorithms that make you buy things when you are reading something on the Web than in the writing that you are reading.

Speech recognition (e.g., Apple's Siri) and image recognition (e.g., Facebook's Deep Face and Microsoft's CaptionBot) will benefit from the progress in neural networks. For example, Apple's Siri, that used speech-recognition technology developed at Nuance before deep learning matured, and that is mainly used to check the weather, will probably benefit from the acquisition of VocalIQ, a spinoff of Cambridge University with experience in deep learning. First demonstrated in 2014, Microsoft's Skype Translate, capable of translating speech in real-time, went live in 2016. In 2016 Google made available Cloud Speech API to the open-source community, so that any developer can power its app with Google's speech recognition.

The next generation of "conversational" agents will be able to access a broader range of information and of apps, and therefore provide the answer to more complicated questions; but they are not conversational at all: they simply query databases and return the result in your language. They add a speech-recognition system and a speech-generation system to the traditional database management system.

There are actually "dream" applications for deep learning. Health care is always at the top of the list because its impact on ordinary people can be significant. The medical world produces millions of images every year: X-Rays, MRIs, Computed Tomography (CT) scans, etc. In 2016 Philips Health Care estimated that it manages 135 billion medical images, and it adds 2 million new images every week. These images are typically viewed by only one physician, the physician who ordered them; and only once. This physician may not realize that the image contains valuable information about something outside the specific disease for which it was ordered. There might be scientific discoveries that affect millions of those images, but there is nobody checking them against the latest scientific announcements. First of all, we would like deep learning to help radiology, cardiology and oncology departments to understand all their images in real time. And then we would like to see the equivalent of a Googlebot (the "crawler" that Google uses to scan all the webpages of the world) for medical images. Imagine a Googlebot for medical images that continuously scans Philips' database and carries out a thorough analysis of each medical image utilizing the latest updates on medical science. Enlitic in San Francisco, Stanford's spinoff Arterys, and Israel's Zebra Medical Vision are the pioneers, but their solutions are very ad-hoc. A medical artificial intelligence would know your laboratory tests of 20 years ago and would know the lab tests of millions of other people, and would be able to draw inferences that no doctor can draw.

In 2015 the USA launched the Precision Medicine Initiative that consists in collecting and studying the genomes of one million people and then matching those genetic data with their health, so that physicians can deliver the right medicines in the right dose to each individual. This project will be virtually impossible without the use of machines that can identify patterns in that vast database.

There are also disturbing applications of the same technology that are likely to spread. The smartphone app FindFace, developed by two Russian kids in their 20s, Artem Kukharenko and Alexander Kabakov, identifies strangers in pictures by searching pictures posted on social media. If you have a presence on social media, the user of something like FindFace can find out who you are by simply taking a picture of you. In 2016 Apple acquired Emotient, a spinoff of UC San Diego, that is working on software to detect your mood based on your facial expression.

An example of unreasonable expectations is Google's self-driving car. The project was launched in 2009 by Sebastian Thrun, the Stanford scientist who had won the DARPA "Grand Challenge" of 2005, a 212-km race across the Nevada desert. Thrun quit in 2013 and was replaced by Chris Urmson, formerly a Carnegie Mellon University student who in 2007 had worked on William Whittaker's victorious team for the DARPA "Urban Challenge" held at George Air Force Base near Los Angeles. (For the record, Chris Urmson left Google in 2016, as had done most of the original team).

The self-driving car may never fully materialize, but the "driver assistant" is coming soon. Mobileye, the Israeli company founded in 1999 that is widely considered the leader in machine-vision technology (and that does not use deep learning) has a much more realistic strategy based on incremental steps to introduce Advanced Driver Assistance Systems (ADAS) that can assist (not replace) drivrs. Otto, founded by one of the engineers who worked on Google's self-driving car, Anthony Levandowski, does not plan to replace the truck driver but to assist the truck driver, especially on long highway drives. Otto, which in 2016 was acquired by Uber, does not plan to build a brand new kind of truck, but to provide a piece of equipment that can be installed on every truck. In 2014 a total of 3,660 people died in the USA in accidents that involved large trucks.

The need for robots is even greater. There are dangerous jobs in construction and steel work that kill thousands of workers every year. According to the International Labor Organization, mining accidents kill more than 10,000 miners every year; and that number does not include all the miners whose life expectancy is greatly reduced by their job conditions.

Robots and drones need eyes to see and avoid obstacles. There will be a market for computer-vision chips that you can install in your home-made drone, and there will be a market for collision-avoidance technology to install in existing cars. Israel's Mobileye and Ireland's Movidius have been selling computer-vision add-ons for machines for more than a decade.

We also need machines to take care of an increasingly elderly population. The combination of rising life expectancy and declining fertility rates is completely reshaping society. The most pressing problem facing humanity as a whole used to be the well-being and the education of children. That was when the median age was 25 or even lower. Ethiopia has a median age of about 19 like most of tropical Africa. Pakistan has a median age of 21. But the median age in Japan and Germany is 46. This means that there are as many people over 46 as there are under 46. Remove teenagers and children: Japan and Germany don't have enough people to take care of people over 46. That number goes up every year. There are more than one million people in Japan who are 90 years old or older, of which 60,000 are centenarians. In 2014, already 18% of the population of the European Union was over 65 years old, almost ten million people. We don't have enough young people to take care of so many elderly people, and it would be economically senseless to use too many young people on such an unproductive task. We need robots to help elderly people do their exercise, to remind them about taking their medicines, to pick up packages at the front door for them, etc.

I am not afraid of robots. I am afraid that robots will not come soon enough.

The robots that we have today can hardly help. Using an IDC report of 2015, we estimated that about 63% of all robots are industrial robots, with robotic assistants (mostly for surgery), military robots and home appliances (like Roomba) sharing the rest in roughly equal slices. The main robot manufacturers, like ABB (Switzerland), Kuka (Germany, being acquired by China's Midea in 2016) and the four big Japanese companies (Fanuc, Yaskawa,

Epson and Kawasaki), are selling mostly or only industrial robots, and not very intelligent ones. Robots that don't work on the assembly line are a rarity. Mobile robots are a rarity. Robots with computer vision are a rarity. Robots with speech recognition are a rarity. In other words, it is virtually impossible today to buy an autonomous robot that can help humans in any significant way other than inside the very controlled environment of the factory or of the warehouse. Nao (developed by Bruno Maisonnier's Aldebaran in France and first released in 2008), RoboThespian (developed by Will Jackson's Engineered Arts in Britain since 2005, and originally designed to be an actor), the open-source iCub (developed by the Italian Institute of Technology and first released in 2008), Pepper (developed by Aldebaran for Japan's SoftBank and first demonstrated in 2014) and the autonomous robots of the Willow Garage "diaspora" (Savioke, Suitable, Simbe, etc) are the vanguard of the "service robot" that can welcome you in a hotel or serve you a meal at the restaurant: "user-friendly" humanoid robots for social interaction, communication and entertainment at public events. In 2016 Knightscope's K5 robot security guard worked in the garage of the Stanford Shopping Center; Savioke's Botlr delivered items to guests at the Aloft hotel in Cupertino; Lowe's superstore in Sunnyvale employed an inventory checker robot built by Bossa Nova Robotics; and Simbe's Tally checked shelves of a Target store in San Francisco. But these are closer to novelty toys than to artificial intelligence. A dog is still a much more useful companion for an elderly person than the most sophisticated robot ever built.

In 1954 Sylvania used robots to publicize its products. Most of today's robots serve the same function: they are cute for taking pictures

The most used robot in the home is iRoomba, a small cylindrical box that vacuums floors. Not exactly the tentacular monster depicted in Hollywood movies. Unfortunately, it will also vacuum money if you drop it on the floor: we cannot trust machines with no common sense, even for the most trivial of tasks.

An industry that stands to benefit greatly from the "rise of the robots" is the toy industry. In 2016 San Francisco-based startup Anki introduced Cozmo, a robot with "character and personality". That's the future of toys, especially in countries like China where the one-child policy has created a generation of lonely children. In fact, we have already been invaded by robots: there are millions of Robosapien robots. The humanoid Robosapien robot was designed by Mark Tilden, a highly respected inventor who used to work at the Los Alamos National Laboratory, and introduced in 2004 by Hong Kong-based WowWee (a company founded in the 1980s by two Canadian immigrants). Most robots will be an evolution of Pinocchio, not of Shakey.

If you consider them robots, the exoskeletons are a success story. These are basically robots that you can wear. The technology was originally developed by the DARPA to help soldiers carry heavy loads, but it is now used to help victims of brain injuries and spinal-cord injuries in several rehabilitation clinics.

ReWalk, founded by an Israeli quadriplegic (Amit Goffer), Ekso Bionics and Suitx (two UC Berkeley spinoffs) and SuperFlex (an

SRI spinoff) already helped paraplegics or seniors walk. Panasonic's ActiveLink has announced an exoskeleton that will help weak nerdy people like me with manual labor that requires physical strength. The cost is still prohibitively high, but one can envision a not-too-distant future in which we will be able to rent an exoskeleton at the hardware story to carry out gardening and home-improvement projects. After you wear it, you can lift weights and hammer with full strength.

In a more distant future, robots may take advantage of projects such as OpenEase, a platform for machines to share knowledge; or RoboHow (2012), that will enable robots to learn new tasks; or RoboBrain (2014), that learns new tasks from human demonstrations and advice.

But first we will need to build robotic arms whose dexterity matches at least the dexterity of a squirrel.

Our hand has dozens of degrees of freedom. Let's say that it has ten (it actually has many more). I can plan the movement of my hand easily ten steps ahead: that's 10 to the 10th to the 10th to... a very huge number. And i can do it without thinking, in a split second. For a robot this is a colossal computational problem. In 2016 Sergey Levine's team at Google Brain trained robots to pick up things that they had never seen before, and to pick up soft and hard objects in different ways. Two groups had already applied deep learning to improving the dexterity of robots: the one led by Abhinav Gupta at Carnegie Mellon University and the other by Ashutosh Saxena (a former pupil of Andrew Ng at Stanford and the brain behind RoboBrain) at Cornell University. But the real issue is dexterity, not deep learning. "High-level reasoning requires very little computation, but low-level sensorimotor skills require enormous computational resources" (Erik Brynjolfsson)

Earlier in the book i mentioned that two of the motivations for doing A.I. were: a business opportunity and the ideal of improving the lives of ordinary people. Both motivations are at work in these projects. Unfortunately, the technology is still primitive. Don't even think for a second that this very limited technology can create an evil race of robots any time soon.

"Nothing in life is to be feared, it is only to be understood" (Marie Curie).

Jobs in the Age of the Robot – Part 5: The Jobs we Need

Asking which jobs will be eliminated by intelligent machines is asking the wrong question. Technology has been replacing humans for a long time, and typically for every job that is destroyed better jobs are created to manage the technology. The question that makes more sense is: which jobs should "not" be replaced by intelligent machines? Which jobs require common sense; which jobs can have disastrous consequences if performed by a person or a machine that doesn't have common sense?

For example, i don't want the self-driving car because a driver without common sense can kill people. The environment has to get a lot more structured to reassure me that intelligent machines (where "intelligent" really means "incredibly stupid") can be trusted with driving a car.

The other question that makes a lot of sense is: which jobs will be done by intelligent machines that today nobody does. Some of these are jobs that we desperately need but that no human is capable of doing, like scanning all those millions of medical images. That machine will not steal anybody's job, and will not risk anybody's life. That machine will simply provide additional information to your doctor about your health, something that today nobody is doing. No job will be lost on account of this machine. On the contrary, several jobs will be created to build this machine, maintain it, update it, and, some day, decommission it and replace it with a better one; and maybe a job will even be created in a museum of old "intelligent" machines; and certainly writers like me will have a job writing about it (or against it).

Over the last 30 years China has built an enormous number of high-rise buildings. Chinese cities need to check the exteriors and the windows of thousands of skyscrapers. It is a crucial task to guarantee safety to millions of people who live and work in those buildings. Alas, we don't have "Spider Men" that can climb the exterior walls of skyscrapers and check every surface and every window. Tiny climbing robots will do that job. They will not steal anybody's job. They will carry out an important task that today nobody is doing. Those robots will need to be built, programmed, maintained, and operated (and marketed, sold, delivered, and, why not, written about in books like this one). Those robots will not kill jobs: they will create millions of jobs all over the world.

Origins of Singularity Thinking

Singularity thinking originated with the essay "Today's Computers, Intelligent Machines and Our Future" (1978) by Hans Moravec of Carnegie Mellon University and with Marvin Minsky's essay "Will Robots Inherit the Earth" (1994), and was popularized by Ray Kurzweil's "The Singularity is Near" (2005) besides Hans Moravec's "Mind Children" (1988) and "Robot - Mere Machine to Transcendent Mind" (1998). David Levy's "Robots Unlimited" (2006) even predicted that machines will soon be conscious. Masahiro Mori, a scientist at the Tokyo Institute of Technology and future president of the Robotics Society of Japan who in 1970 had published the influential article "The Uncanny Valley", had actually predated the whole Singularity movement when he argued in "The Buddha in the Robot" (1974) that robots would someday be able to attain buddhahood.

Trivia: Daniel Wilson wrote a hilarious manual to help humans survive in a world threatened by intelligent machines, "How to Survive a Robot Uprising" (2005).

The original prophet of what came to be called "transhumanism" was probably Fereidoun "FM-2030" Esfandiary who wrote "Are You a Transhuman?" (1989) and predicted that "in 2030 we will be ageless and everyone will have an excellent chance to live forever". He died of pancreatic cancer (but was promptly placed in cryogenic suspension).

In 1999 Kurzweil argued that there exists a general law, the "Law of Accelerating Returns" that transcends Moore's Law. Order causes more order to be created and at a faster rate. Order started growing exponentially millions of years ago, and progress is now visible on a daily basis. This echoed science-fiction writer Vernon Vinge's declaration that "the acceleration of technological progress has been the central feature of this century" (1993). They both base their conclusions on the ever more frequent news of technological achievements. (Personally, i think that they are confusing progress and the news cycle. Yes, we get a lot more news from a lot more sources. If the same news and communication tools had been available at any time in previous peacetime periods, the people alive back then would have been flooded by an equal amount of news). In particular, at some point computers will acquire the ability to improve themselves, and then the process that has been manually done by humans will be automated like many other manual jobs, except that this one is about making smarter computers, which means that the process of making smarter computers will be automated by smarter computers, which turns into a self-propelled accelerating loop. This will lead to an infinite expansion of "intelligence".

The Case for Superhuman Intelligence... and against it

The case for the coming of an artificial intelligence, of an artificial general intelligence and then of the Singularity rests on the simple assumption that A.I. is making dramatic progress and that progress is accelerating. If you believe these two statements, then you probably believe that we will soon have machines that can have a philosophical conversation with us and write books like this one.

That is precisely the conclusion that Moravec and Kurzweil reached. Hans Moravec, the author of "Mind Children" (1988) and "Robot - Mere Machine to Transcendent Mind" (1998), predicted that machines will become smarter than humans by 2050. Ray Kurzweil, the author of "The Singularity is Near" (2005), predicted that machine intelligence will surpass human intelligence by 2045.

Moravec and Kurzweil were not the first futurists to put those two assumptions together. In 1957 Herbert Simon, one of the founders of A.I., had said: "there are now in the world machines that think, that learn, and that create. Moreover, their ability to do these things is going to increase rapidly".

The case against A.I. (and therefore the Singularity) dates from the 1970s, when philosophers started looking into the ambitious statements coming out of the A.I. world.

The first philosopher to look into these claims was Hubert Dreyfus, who wrote in "Alchemy and Artificial Intelligence" (1965): "Significant developments in Artificial Intelligence … must await an entirely different sort of computer. The only existing prototype for it is the little-understood human brain."

Mortimer Taube, author of "Computers and Common Sense" (1961), and John Lucas, author of "Minds, Machines and Gödel" (1961), had already pointed out that full machine intelligence is incompatible with Kurt Gödel's incompleteness theorem. In 1935 Alonso Church proved a theorem, basically an extension of Gödel's incompleteness theorem to computation: that first-order logic is "undecidable". Similarly, in 1936 Alan Turing proved that the "halting problem" is undecidable for Universal Turing Machines. What these two theorems say is basically that it cannot be proven whether a computer will always find a solution to every problem; and that is a consequence of Gödel's theorem, a highly respected mathematical proof. Several thinkers have used similar arguments based on Gödel's theorem, notably Roger Penrose in "The Emperor's New Mind" (1989).

The most famous critique of Artificial Intelligence was contained in John Searle's article "Minds, Brains and Programs" (1980) and came to be known as the "Chinese room" argument. If you give a person a comprehensive set of instructions needed to translate Chinese into English and lock that person in a room, someone standing outside the room would be fooled into thinking that the person inside knows Chinese, when in fact that person is mechanically following instructions that are meaningless to her to manipulate symbols that are also meaningless. She has no clue what the Chinese sentence says, but she produces the correct translation into English. Endless papers have been written by philosophers to discuss the validity of Searle's argument. However, Searle was not attacking the feasibility of machine intelligence but simply whether an intelligent machine would be also conscious.

Today's computers, including the superfast GPUs used by AlphaGo, are Turing Machines. Critics of A.I. need to prove that Turing machines cannot match human intelligence. Many have written books along those lines, but what are the tasks that Turing Machines can't perform is a conveniently movable target. To my knowledge, nobody spelled out what is it that Turing machines will never do better than us.

I suspect that here another kind of "religion" plays a role in the opposite direction, in the direction of making us reluctant to accept that machines can become as intelligent as us and even more intelligent. Astrophysics has shown that there is nothing special about the location of the Earth, Biology showed that there is nothing special about human life, neuroscience is showing that there is nothing special about the human brain, and now Artificial Intelligence might show that there is nothing special about our intelligence. Each of these revelations seems to make humankind less relevant, more insignificant.

But i have seen no convincing proof that machines can reach human-level intelligence. Hence: why not?

Instead, one has to wonder for how long it will make sense to ask the question whether full-fledged artificial intelligence is possible. If the timeframe for fully intelligent machines is centuries and not decades like the optimists believe, then it's like asking an astronaut

"Will it at some point be possible to send a manned spaceship to Pluto?" Yes, it may be very possible, but it may never happen: not because it's impossible but simply because we may invent teleportation that will make spaceships irrelevant. Before we invent intelligent machines, synthetic biology or some other discipline might have invented something that will make robots irrelevant. The timeframe is not a detail.

Assuming that some day we will have fully intelligent machines, will they evolve into a superior level of intelligence that is unattainable by humans? That is a different question. I have seen no proof that machine intelligence inevitably leads to machines becoming more intelligent than humans.

Let me use a metaphor. Just because we built a ladder it doesn't mean that we can fly: it only means that we can build taller and taller ladders, and maybe those ladders will help us climb on the roof and fix a leak; but the technology to fly is different from the technology of climbing ladders, and therefore virtually no progress towards flying will be achieved by building better and better ladders. And I doubt that ladders will spontaneously evolve into flying beings. Both the ladder and the bird have to do with "heights" and naive media may conclude that one leads to the other, but people who build ladders should know better.

What exactly are the things that a superhuman intelligence can do and no human being can ever do? If the answer is "we cannot even conceive them", then we are back to the belief that angels exist and miracles happen, something that eventually gave rise to organized religions. If instead there is a simple, rational definition of what a superhuman intelligence can do that no human can ever do, i have not seen it; or, better, i have seen one but also the opposing view. I will briefly discuss these two opposite views.

On one hand, superhuman intelligence should exist because of the "cognitive closure", a concept popularized by Colin McGinn in "The Problem Of Consciousness" (1991). The general idea is that every cognitive system (e.g., every living being) has a "cognitive closure": a limit to what it can know. A fly or a snake cannot see the world the way we see it because they do not have the same visual system that humans have. In turn, we can never know how it feels to be a fly or a snake. A blind person can never know what "red" is even after studying everything that there is to be studied about the color "red". According to this idea, each brain (including the human brain) has a limit to what it can possibly think, understand, and know. In particular, the human brain has a limit that will preclude humans from understanding some of the ultimate truths of the universe. These may include spacetime, the meaning of life, and consciousness itself. There is a limit to how "intelligent" we humans can be. According to this view, there should exist cognitive systems that are "superhuman", i.e. they don't have the limitations that our cognition has.

However, i am not sure if we (humans) can intentionally build a cognitive system whose cognitive closure is larger than ours, i.e., a cognitive system that can "think" concepts that we cannot think. It sounds a bit of an oxymoron that a lower form of intelligence can intentionally build the highest form of intelligence. However, it is not a contradiction that a lower form of intelligence can accidentally (by sheer luck) create a higher form of intelligence.

That is the argument in favor of the feasibility of superhuman intelligence. A brilliant argument against such feasibility is indirectly presented in David Deutsch's "The Beginning of Infinity" (2011). Deutsch argues that there is nothing in our universe that the human mind cannot understand, as long as the universe is driven by universal laws. I tend to agree with Colin McGinn that there is a "cognitive closure" for any kind of brain, that any kind of brain can only do certain things, and that our cognitive closure will keep us from ever understanding some things about the world (perhaps the nature of consciousness is one of them); but in general i also agree with Deutsch: if something can be expressed in formulas, then we humans will eventually "discover" it and "understand" it; and, if everything in nature can be expressed in formulas, then we (intelligent beings) will eventually "understand" everything, i.e. we are the highest form of intelligence that can possibly exist. So the only superhuman machine that would be too intelligent for humans to understand is a machine that does not obey the laws of nature, i.e. that is not a machine.

If you lean towards the "cognitive closure" argument, you also have to show that we haven't reached it yet. The progress of the human mind did not necessarily end with you. If human intelligence hasn't reached the cognitive closure yet, then there is still room for improvement in human intelligence. I see no evidence that the human mind may have reached a maximum of creativity and will never go any further. We build machines based on today's knowledge and creativity. Maybe, some day, those machines will be able to do everything that we do today; but why should we assume that, by then, the human mind will not have progressed to new levels of knowledge and creativity? By then, humans may be thinking in different ways and may invent things of a different kind. Today's electronic machines may continue to exist and evolve for a while, just like windmills existed and evolved and did a much better job than humans at what they were doing; but some day electronic machines may look as archaic as windmills look today. I suspect that there is still a long way to go for human creativity. The Singularity crowd cannot imagine the future of human intelligence the way that someone in 1904 could not imagine Relativity and Quantum Mechanics.

Some day the Singularity might come, but i wouldn't panic. Mono-cellular organisms were neither destroyed nor marginalized by the advent of multicellular organisms. Bacteria are still around, and probably more numerous than any other form of life in our part of the universe. The forms of life that came after bacteria were presumably inconceivable by bacteria but, precisely because they were on a different plane, they hardly interact. We kill bacteria when they harm us but we also rely on many of them to work for us (our body has more bacterial cells than human cells). In fact, some argue that a superhuman intelligence already exists, and it's the planet as a whole, Gaia, of which we are just one of the many components.

In some cases we are "afraid" of a machine simply because we can't imagine the consequences. Imagine the day when machines will be able to understand natural language. A human can read only a few books a week. Such a machine, instead, will be able to read in a few seconds all the texts ever produced and digitized by the human race. It is hard to imagine what this implies.

In theory, an artificial intelligence that talks to another artificial intelligence could learn a lot faster than us. We humans need to relocate ourselves to places called universities and take lengthy classes to learn just a fraction of what the experts know. An artificial intelligence could learn in just a few seconds everything that another artificial intelligence knows (with a single "memory dump"). In fact, some day (if computer speed keeps improving) an artificial intelligence could learn everything that EVERY artificial intelligence knows. Imagine if you could learn in a few seconds everything that all humans know.

The way our bodies and brains are built by nature makes it impossible for us to do the same. One possibility is that Nature couldn't do any better. The other possibility is that, maybe, over millions of years of natural selection, Nature figured out that it is better that way.

Critics of A.I. cannot tell us exactly what it is that machines will never be able to do that humans can do. Believers in the Singularity cannot tell us exactly what it is that humans will never be able to do that machines will do. My tentative conclusion is that machines as intelligent as humans are possible (the question is not "if" but "when") whereas machines more intelligent than humans are not possible. Alas, this conclusion hinges on a very vague definition of "intelligence".

What is the Opposite of the Singularity?

What worries me most is not the rapid increase in machine intelligence but a possible decrease in human intelligence.

The Turing Test is commonly understood as: when can we say that a machine has become as intelligent as humans? But the Turing Test is about humans as much as it is about machines because it can equivalently be formulated as: when can we say that humans have become as stupid as a machine? In other words, there is another way for machines to pass the Turing Test: make dumber humans. Let's call the Turing Point the point when the machine has become as smart as humans. The Turing Point can be reached because machine intelligence increases to human level or because human intelligence decreases to machine level.

Humans have always become dependent on the tools they invented. For example, when they invented writing, they lost memory skills. On the other hand, they discovered a way to store a lot more knowledge and to disseminate it a lot faster. Ditto for all other inventions in history: a skill was lost, a skill was acquired. We cannot replay history backwards and we will never know what the world would be like if humans had not lost those memory skills. Indirectly we assume that the world as it is now is the best that it could have been. In reality, over the centuries the weaker memory skills have been driving an explosion of tools to deal with weak memory. Each tool, in turn, caused the decline of another skill. It is debatable if the invention of writing was worth this long chain of lost skills. This process of "dumbification" has been going on throughout society and it accelerated dramatically and explosively with electrical appliances and now with digital devices. The computer caused the decline of calligraphy. Voice recognition will cause the decline of writing.

In a sense, technology is about giving dumb people the tools to become dumber and still continue to live a happy life. A pessimist can rewrite the entire history of human civilization as the history of making humans dumber and inventing increasingly smarter tools to compensate for their increasing stupidity.

In some cases the skill that is lost may have broader implications. If you always use the smartphone's navigator to find places, your brain does not exercise the part of the brain that knows how to navigate the territory. If we don't explore, we don't learn how to explore. If we don't learn how to explore, we don't grow cognitive maps, and we don't train the brain to create cognitive maps. The cognitive map is a concept introduced in 1948 by Edward Tolman to explain how higher animals orient themselves. Without them, the brain is a diminished organ: it will never learn how to do all the things that are enabled by cognitive maps. If George Lakoff is right and all thinking is rooted in physical metaphors, there are countless thoughts that can happen only to brains that know how to manage cognitive maps. Reading novels and discovering scientific theories may not be possible without cognitive maps.

What can machines do now that they could not do 50 years ago? They are just faster, cheaper and can store larger amounts of information. These factors made them ubiquitous. What could humans do 50 years ago that they cannot do now? Ask your grandparents and the list is very long, from multiplication to orientation, from driving in chaotic traffic to fixing a broken shoe. Or just travel to an underdeveloped country where people still live like your old folks used to live and you will find out how incapable you are of performing simple actions that are routine for them. When will we see a robot that is capable of crossing a street with no help from the traffic light? It will probably take several decades. When will we get to the point that the average person is no longer capable of crossing a street without help from the traffic light? That day is coming much sooner. Judging from simple daily chores, one could conclude that human intelligence is not "exploding" but imploding. Based on the evidence, one can argue that machines are not getting much smarter (just faster), while humans are getting dumber; hence very soon we will have machines that are smarter than humans but not only because machines got smarter.

The age of digital devices is enabling the average person to have all sorts of knowledge at her fingertips. That knowledge originally came from someone who was "intelligent" in whichever field. Now it can be used by just about anybody who is not "intelligent" in that field. This user has no motivation to actually learn it: she can just "use" somebody else's "intelligence". The "intelligence" of the user decreases, not increases (except, of course, for the intelligence on how to operate the devices; but, as devices become easier and easier to use, eventually the only intelligence required will be to press a button to turn the device on). Inevitably, humans are becoming ever more dependent on machines, while machines are becoming less dependent on humans.

I chair/organize/moderate cultural events in the Bay Area and, having been around in the old days of the overhead projectors, i'm incredulous when a speaker cannot give her/his talk because her/his computer does not connect properly to the room's audio/visual equipment and therefore s/he cannot use the prepared slide presentation. For thousands of years humans were perfectly capable of giving a talk without any help from technology. Not anymore, apparently. Can you imagine Socrates telling Plato "Sorry, I can't have a dialogue with you unless you have Powerpoint on your laptop"?

The Turing Test could be a self-fulfilling prophecy: at the same time that we (claim to) build "smarter" machines, we are creating dumber people.

My concern, again, is not for machines that are becoming too intelligent, but for humans who are becoming less intelligent. What might be accelerating is the loss of human skills. Every tool deprives humans of the training they need to maintain a skill (whether arithmetic or orientation) and every interaction with machines requires humans to lower their intelligence to the intelligence of machines (e.g., to press digits on a phone in order to request a service). We can argue forever if the onboard computer of a self-driving car is really "driving", but we know for sure what the effect of self-driving cars will be: raising a generation of humans that is incapable of driving anymore. Every machine that replaces a human skill (whether the pocket calculator or the street navigator) reduces the training that humans get in performing that skill (such as arithmetic and orientation), and eventually causes humans to lose that skill. This is an ongoing experiment on the human race that could have a spectacular result: the first major regression in intelligence in the history of our species.

To be fair, it is not technology per se that makes us dumber. The very system that produces technology makes us dumber. The first step usually consists in some rules and regulations that simplify and normalize a process, whether serving food at a fast-food chain or inquiring about the balance of your checking account or driving a car. Once those rules and regulations are in place, it gets much easier to replace human skills with technology: the human skills required to perform those tasks have been reduced dramatically, and, in that sense, humans have become "dumber" at those tasks. In a sense, technology is often an effect, not a cause: once the skills required to perform a task have been greatly downgraded, it is quite natural to replace the human operator with a machine.

Paraphrasing something Bertrand Russell said about Ludwig Wittgenstein, we are weary of thinking and we are building a society that would make such an activity unnecessary. Then, of course, an unthinking machine would equal an unthinking human, not because the machine has become as thinking as the human, but because the human has become as unthinking as the machine.

The society of rules and regulations that humans have built to create order and stability has the side effect of making us "think" less.

The Turing Test can be achieved in two ways: 1. by making machines so intelligent that they will seem human; 2. by making humans so stupid that they will seem mechanical.

To wit, there could be three stages in human civilization. Stage 1: the coexistence of machine stupidity and human intelligence. Stage 2: the coexistence of machine intelligence and human intelligence. Stage 3: the coexistence of machine intelligence and human stupidity.

With all due respect, when i interact with government officials or corporate employees, the idea that these people, trained like monkeys to repeatedly say and do the prescribed routine, will some day be enslaved by intelligent machines does not seem so implausible.

What will "singular" mean in a post-literate and post-arithmetic world?

"Men have become the tools of their tools" (Henry Thoreau, "Walden", 1854).

Intermezzo: The Attention Span

This topic has more to do with modern life than with machines, but it is related to the idea of an "intelligence implosion".

I worry that the chronic scarcity of time in our age is pushing too many decision makers to take decisions having heard only very superficial arguments. The "elevator pitch" has become common even in academia. A meeting that lasts more than 30 minutes is a rarity (in fact, a luxury from the point of view of the most powerful, and therefore busiest, executives). You can't get anybody's attention for more than 20 minutes, but some issues cannot be fully understood in 20 minutes; and some great scientists are not as good at rhetorical speech as they are at their science, which means that they may lose a 20-minute argument even if they are 100% right. Too many discussions are downgraded because they take place by texting on so-called smartphones, whose tiny keyboards discourage elaborate messages. The ultimate reason that we have fewer and fewer investigative reporters in news organizations is the same, i.e. the reduced attention span of the readers/viewers, with the result that the reliability of news media is constantly declining. Twitter's 140-character posts have been emblematic of the shrinking attention span.

(Trivia: Twitter introduced the limitation of 140 characters on human intelligence in the same year, 2006, that deep learning increased the intelligence of machines).

I am not afraid that the human race might lose control of its machines as much as i am afraid that the human race will self-destruct because of the limitations of the "elevator pitch" and of the "tweet"; because of the chronic inability of decision makers, as well as of the general public, to fully understand an issue.

It has become impossible to properly organize events because the participants, accustomed to tweets and texting, will only read the first few lines of a lengthy email. Multiply this concept a few billion times in order to adapt it to the dimension of humanity's major problems, and you should understand why the last of my concerns is that machines may become too intelligent and the first of my concerns is that human interactions might become too dumb. Elon Musk (at MIT's AeroAstro 100 conference in October 2014) and others are worried that machines may get so smart that they will start building smarter machines; instead, i am worried that people's attention span is becoming so short that it will soon be impossible to explain the consequences of a short attention span. I don't see an acceleration in machine intelligence, but i do see a deceleration in human attention… if not in human intelligence in general.

To summarize, there are three ways that we can produce "dumber" humans. All three are related to technology but in opposite ways.

Firstly, there is the simple fact that a new technology makes some skills irrelevant, and those skills may be lost within one generation. Pessimists argue that little by little we become less human. Optimists claim that the same technology enables new skills to develop. I can personally attest that both camps are right: the computer and email have turned me into a highly-productive multi-tasking cyborg, and at the same time they have greatly reduced my skills in writing polite and touching letters to friends and relatives (with, alas, parallel effects on the quality of my poetry). The pessimists think that the gains do not offset the losses (the "dumbification"), especially when it comes to losing basic survival skills.

Secondly, the rules and regulations that society introduces for the purpose of making us safer and more efficient end up making us think less and less, i.e. behave more and more like (non-intelligent) machines.

Thirdly, the frantic lives of overworked individuals have greatly reduced their attention span, which may result in a chronic inability to engage in serious discussions, i.e. in a more and more superficial concept of "intelligence", i.e. in the limited cognitive experience of lower forms of life.

Cognitive Intermezzo: The Origin of Human Intelligence (or of Machine Intelligence?)

Steven Piantadosi and Celeste Kidd at Rochester University ("Extraordinary Intelligence and the Care of Infants" in Proceedings of the National Academies of Science, 2016) have found a correlation in primates between the degree of intelligence in the adults and the degree of "stupidity" in their offspring. Human brains are so sophisticated because human parents need to take care of the most helpless babies in the animal kingdom. The dumber the children, the smarter the parent has to be to keep them alive. Other animals start walking and eating right after being born. Human babies need to be fed and learn to walk only after many months. Their theory is: the dumber the children, the smarter the parents must be. How do you create very intelligent parents? By giving them very dumb children to watch, protect, lecture, etc. They speculate that there is a self-reinforcing loop at work: because humans make the dumbest children, they must be very intelligent adults, and in order to produce intelligent adults the children must be very dumb. I wonder if a similar self-reinforcing loop is at work on the intelligence of technology: does technology make dumber humans in order for humans to invent smarter technology to deal with dumber humans, technology which will in turn make humans even dumber?

Anthropological Intermezzo: You Are a Gadget

The combination of phones, computers and networks has put each individual in touch with a great number of other individuals, more than at any time in history: humankind at your fingertips. This is certainly lucrative for businesses that want to reach as many consumers as possible with their ads. But do ordinary people really benefit from being connected to thousands of people, and soon millions? What happens to solitude, meditation, to "thinking" in general (whether scientific thinking or personal recollection) when we are constantly interacting with a multitude of minds (only some of which really care)?

You "are" the people with whom you interact, because they influence who you become. In the old days those were friends, relatives, neighbors and coworkers. Now they are strangers spread all over the world (and old acquaintances with whom you only share distant memories). Do you really want to be "them" rather than being yourself?

You are not surrounding yourself with people: you are surrounding yourself with gadgets like smartphones and laptops.

If you surround yourself with philosophers, you are likely to become a philosopher, even if only an amateur one. If you surround yourself with book readers, you are likely to read a lot of books. If you surround yourself with physicists, you are likely to understand Relativity and Quantum Mechanics. And so on. So what is likely to happen to you if you surround yourself with gadgets that mediate your interaction with people and with the world at large?

It is infinitely easier to produce/accumulate information than to understand it and make others understand it.

Existential Intermezzo: You Are an Ad

We are surrounded by billboards, ads and commercials. Even the webpages that you visit on the Web depend on an algorithm of page-ranking whose behavior can be manipulated by expert professionals, so that you will visit the webpages that they want you to visit and not the ones that you would have visited using brain and luck. If you search for "Piero" with your favorite search engine, you are more likely to stay at an apartment complex in Los Angeles and eat at a restaurant in Las Vegas than read a page on my website, a website that has been around for 20 years, contains 10,000 pages of text, and is edited by a writer named Piero.

In the 1990s my website would show up as one of the top three results of a search for "Piero" on every search engine. I'll let you decide if the results that you get today are "better" than what you were getting 20 years ago. You have to thank progress in Artificial Intelligence for it.

Artificial Intelligence is making these ads more powerful, more targeted, more convincing, more inescapable.

Our lives are increasingly steered by the advertisement that surrounds (traps?) us.

What happens to us if (when) the ads stop?

Will we still able to live a life?

There might be a sinister new meaning for Eliot's famous line "You are the music while the music lasts" (T. S. Eliot, "The Dry Salvages", 1941)

Semantics

In private conversations about "machine intelligence" i like to quip that it is not intelligent to talk about intelligent machines: whatever they do is not what we do, and, therefore, is neither "intelligent" nor "stupid" (attributes invented to define human behavior). Talking about the intelligence of a machine is like talking about the leaves of a person: trees have leaves, people don't. "Intelligence" and "stupidity" are not properties of machines: they are properties of humans. Machines don't think, they do something else. Machine intelligence is as much an oxymoron as human furniture. Machines have a life of their own, but that "life" is not human life.

We apply to machines many words invented for humans simply because we don't have a vocabulary for the states of machines. For example, we buy "memory" for our computer, but that is not memory at all: it doesn't remember (it simply stores) and it doesn't even forget, the two defining properties of (biological) memory. We call it "memory" for lack of a better word. We talk about the "speed" of a processor but it is not the "speed" at which a human being runs or drives. We don't have the vocabulary for machine behavior. We borrow words from the vocabulary of human behavior. It is a mistake to assume that, because we use the same word to name them, they are the same thing. If i see a new kind of fruit and call it "cherry" because there is no word in my language for it, it doesn't mean it is a cherry. A computer does not "learn": what it does when it refines its data representation is something else (that we don't do).

It is not just semantics. Data storage is not memory. Announcements of exponentially increasing data storage miss the point: that statistical fact is as relevant to intelligence as the exponential increase in credit card debt. Just because a certain sequence of zeroes and ones happens to match a sequence of zeroes and ones from the past it does not mean that the machine "remembered" something. Remembering implies a lot more than simply finding a match in data storage. Memory does not store data. In fact, you cannot retell a story accurately (without missing and possibly distorting tons of details) and you cannot retell it twice with the same words (each time you will use slightly different words). Ask someone what her job is, something that she has been asked a thousand times, and she'll answer the question every time with a different sequence of words, even if she tries to use the same words she used five minutes earlier. Memory is "reconstructive", the crucial insight that Frederic Bartlett had in 1932. We memorize events in a very convoluted manner, and we retrieve them in an equally convoluted manner. We don't just "remember" one thing: we remember our entire life whenever we remember something. It's all tangled together. You understand something not when you repeat it word by word like a parrot (parrots can do that, and tape recorders can do that) but when you summarize it in your own words, different words than the ones you read or heard: that is what we call "intelligence". I am always fascinated, when i write something, to read how readers rewrite it in their own words, sometimes using completely different words, and sometimes saying it better than i did.

It is incredible how bad our memory is. A friend recommended an article by David Carr about Silicon Valley published in the New York Times "a few weeks ago". It took several email interactions to figure out that a) the article was written by George Packer, b) it was published by the New Yorker, c) it came out one year earlier. And, still, it is amazing how good our memory is: it took only a few sentences during a casual conversation for my friend to relate my views on the culture of Silicon Valley to an article that she had read one year earlier. Her memory has more than just a summary of that article: it has a virtually infinite number of attributes linked to that article such that she can find relevant commonalities with the handful of sentences she heard from me. It took her a split second to make the connection between some sentences of mine (presumably ungrammatical and incoherent sentences because we were in a coffee house and i wasn't really trying to compose a speech) and one of the thousands of articles that she has read in her life.

All forms of intelligence that we have found so far use memory, not data storage. I suspect that, in order to build an artificial intelligence that can compete with the simplest living organism, we will first need to create artificial memory (not data storage). Data storage alone will never get you there, no matter how many terabytes it will pack in a millimeter.

What computers do is called "savant syndrome" in the scientific literature: idiots (very low intelligence quotient) with a prodigious memory.

I am not advocating that machines should be as forgetful and slow as us. I am simply saying that we shouldn't be carried away by a faulty and misleading vocabulary.

Data is not knowledge either: having amassed all the data about the human genome does not mean that we know how human genes work. We know a tiny fraction of what they do even though we have the complete data.

I was asking a friend how the self-driving car works in heavy traffic and he said "the car knows which other cars are around". I object that the car does not "know" it. There is a system of sensors that continuously relay information to a computer that, in turn, calculates the trajectory and feeds it into the motor controlling the steering wheel. This is not what we mean when we say that we "know" something. The car does not "know" that there are other cars around, and it does not "know" that cars exist, and it doesn't even "know" that it is a car. It is certainly doing something, but can we call it "knowing". This is not just semantics: because the car does not "know" that it is a car surrounded by other cars driving on a road, it also lacks all the common sense or general knowledge that comes with that knowledge. If an elephant fell from the sky, a human driver would be at least surprised (and probably worried about stranger phenomena ahead), whereas a car would simply interpret it as an object parked in the middle of the highway.

When, in 2013, Stanford researchers trained a robot to take the elevator, they realized that there was a non-trivial problem: the robot stopped in front of the glass doors of the elevator interpreting its own reflection into it as another robot. The robot does not "know" that the thing is a glass door otherwise it would easily realize that there is no approaching robot, just a reflection getting bigger like all reflections do when you walk towards a mirror.

It is easy to claim that, thanks to Moore's law, today's computers are one million times faster than the computers of the 1980s, and that a smartphone is thousands of times faster than the fastest computer of the 1960s. But faster at what? Your smartphone is still slower than a snail at walking (it doesn't move, does it?) and slower than television at streaming videos. Plenty of million-year old artifacts are faster than the fastest computer at all sorts of biological processes. And plenty of analog devices (like television) are still faster than digital devices at what they do. Even in the unlikely event that Moore's law applies to the next 20 years, processing speed and storage capacity will improve by a factor of a million. But that may not increase at all the speed at which a computer can summarize a film. The movie player that i use today on my laptop is slower (and a lot less accurate) in rewinding a few scenes of the film than the old videotape player of twenty years ago, no matter how "fast" the processor of my laptop is.

We tend to use cognitive terms only for machines that include a computer, and this habit started way back when computers were invented (the "electronic brains"!). Thus the cognitive vocabulary tempts people to attribute "states of mind" to those machines. We don't usually do this to other machines. A washing machine washes clothes. If a washing machine is introduced that washes tons of clothes in a split second, consumers will be ecstatic, but presumably nobody would take it as an example of human or superhuman intelligence. And note that appliances do some pretty amazing things. There's even a machine called "television set" that shows you what is happening somewhere else, a feat that no intelligent being can do. We don't attribute cognitive states to a television set even though the television set can do something that requires more than human intelligence.

Take happiness instead of intelligence. One of the fundamental states of human beings is "happiness". When is a machine "happy"? The question is meaningless: it's like asking when does a human being need to be watered? You water plants, not humans. Happiness is a meaningless word for machines. Some day we may start using the word "happy" to mean, for example, that the machine has achieved its goal or that it has enough electricity; but it would simply be a linguistic expedient. The fact that we may call it "happiness" does not mean that it "is" happiness. If you call me Peter because you can't spell my name, it does not mean that my name is Peter.

Semantics is important to understand what robots really do. Pieter Abbeel at UC Berkeley has a fantastic robotic arm that can fold towels with super-human dexterity. But what it does is "not" what a human does when she folds towels. Abbeel's robot picks up a towel, shakes it, turns, and folds it on a table. And it does it over and over again, implacably, without erring. What any maid in a hotel does is different. She picks up a towel and folds it... unless the towel is still wet, or has a hole, or needs to be washed again, or... That is what "folding towels" really means. The robot is not folding towels: it is performing a mechanical movement that results in folded towels, some of which may be entirely useless for human purposes.The maid is not paid to do what the robot does. She is paid to fold towels, not to "fold towels" (the quotes make the difference).

Cleaning up a table at a restaurant is not just about throwing away all the objects that are lying on the table, but about recognizing which objects are garbage and which are not (eg a crumpled piece of paper versus paper money), which ones must be moved elsewhere and which ones belong there (e.g. a vase full of flowers, but not a vase full of withered flowers). A cell phone left on the table is neither garbage nor a dish to be washed: it is something that the customer forgot behind.

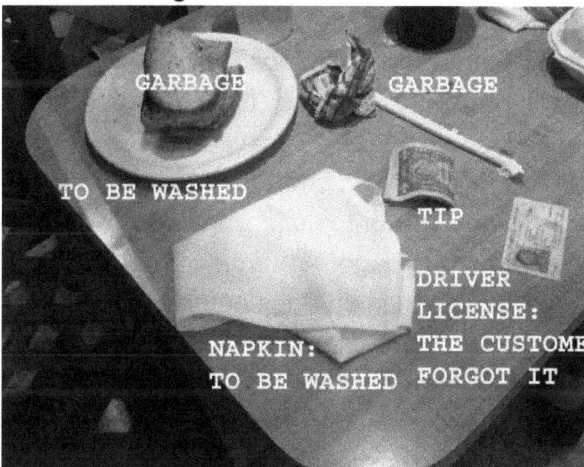

It is true that machines can now recognize faces, and even scenes, but they have no clue what those scenes mean. We will soon have machines that can recognize the scene "someone picked up an object in a store", but when will we have a machine that can recognize "someone STOLE an object from a store?" A human being understands the meaning of this sentence because humans understand the context: some of those objects are for sale, or belong to the shop, and a person walking away with those objects is a thief, which is very different from being a store clerk arranging the goods on the shelves or a customer bringing the object to the counter and paying for it. We can train neural networks to recognize a lot of things, but not to understand what those things mean.

And the day when we manage to build a machine that can recognize that someone is stealing someone else's wallet, we will still have a higher level of understanding to analyze: that scene could be a prank, as indicated by the fact that one of the two people is smiling, and that we know that they are old friends. In that case you don't call the police but simply wait for the fun to begin. And if we build a machine that can even recognize a prank, we will still have go up one more level of abstraction to consider the case in which this is happening in a movie, not in reality. And so on and so forth. The human mind can easily grasp these situations: the same scene can mean so many different things.

The automatic translation software that you use to translate Chinese into English doesn't have a clue what those Chinese words mean nor what those English words mean. If the sentence says "Oh my god there's a bomb!" the automatic translation software simply translates it into another language. A human interpreter would shout "everybody get out!", call the emergency number and... run!

Intelligence is not about the error rate in recognizing an action. Intelligence is about "recognizing" the action for what it is. Mistakes are actually fine. We make mistakes all the time. Sometimes we think we recognized an old friend and instead it turns out to be a complete stranger. We laugh and move on. And that's another way to ponder the difference in semantics. We laugh out loud when we see the kind of mistakes that a computer makes when it is trying to recognize a scene. I just searched for images related to "St Augustine what is time" and the most famous search engine returned a page of pizza images. This is what humans do: humans laugh out loud when someone (or something) makes such silly mistakes. The real Turing Test is this: when will we have a computer that laughs out loud at silly mistakes made by other computers or by itself?

Using human semantics, the most intelligent machines ever built, such as IBM's "Watson" and Google's "AlphaGo", are incredibly stupid. They can't even cook an omelette, they cannot sort out my clothes in the drawers, they cannot sit on the sidewalk and gossip about the neighborhood (i am thinking of human activities that we normally don't consider "very intelligent"). A very dumb human being can do a lot more than the smartest machines ever built, and that's probably because there is a fundamental misunderstanding about what "intelligent" means.

AlphaGo did not "win" a game of weichi/go. AlphaGo never learned how to play weichi. AlphaGo cannot answer any question, but, if it could, it would not know the answer to the question "what are the rules of weichi?" AlphaGo never learned to play weichi. AlphaGo simply calculates the most likely good move based on moves made by thousands of go masters in similar situations. AlphaGo has no idea that it is playing go, playing a game, playing with humans, etc. Therefore it does not "win" or "lose".

There is also talk of "evolving A.I. systems", which projects the image of machines getting more and more intelligent. This can mean different things: a) a software that devises a better technique to solve problems; b) a software that has improved itself through learning from human behavior; c) a software that has improved itself through self-playing. None of this is what we mean when we say that a species evolved in nature. Evolution in nature means that a population makes children that are all slightly different and then natural selection rewards the ones that are the best fit to the environment. After thousands of generations, the population will evolve into a different species that will not mate with the original one. There is nothing wrong with software programs that get better at doing what they do, but calling it "evolution" evokes a metaphor (and an emotional reaction) that just does not apply to today's software. There is no software that evolves. And even if you really want to call it "evolution", you should realize that the software program has "evolved" because of the software engineer who programmed it. If tomorrow beavers start building better dams, do you talk about the evolution of dams or the evolution of beavers?

The mother of all misunderstandings is the fact that we classify some technologies under the general label "Artificial Intelligence", which automatically implies that machines equipped with those technologies will soon become as intelligent as humans. There are many technologies that have made, are making and will make machines more intelligent. For example, the escapement and the gyroscope made several machines more intelligent, from clocks to motion-sensing devices, but people are not alarmed that escapements and gyroscopes might take over the world and kill us all. Monte Carlo methods have been widely used in simulation since Stanislaw Ulam published the first paper in 1949. They are usually classified under "Numerical Analysis" and sometimes under "Statistical Analysis", and don't scare anybody. Mathematically speaking, they apply statistical methods to find a solution to problems that are described by mathematical functions with no known solution. Sounds boring, right? But the Monte-Carlo tree search is a Monte Carlo method used by AlphaGo for determining the best move in a game. Now it doesn't sound boring anymore, right? If we now classify the Monte Carlo method under "Artificial Intelligence", we suddenly turn a harmless statistical technique into some kind of dangerous intelligent agent, and the media will start writing articles about how this technique will create super-intelligent machines. That is precisely what happened with "neural networks". When in 1958 the psychologist Frank Rosenblatt built the first "neural network", his aim was indeed to model how the human brain works. Today we know that the similarities are vague at best. It is like comparing a car to a horse because the car was originally called the "horseless carriage" (we still measure a car's power in horsepower!) Progress in neural networks has not been based on neuroscience but on computational mathematics: we need mathematical functions that can be implemented in computers and that can yield solutions in a finite time. Calling them "neural networks" makes people think of brains, and turns them into ideas for Hollywood movies. If we called them "constraint propagation" (which is what they are), they would only make people think of the algebra they hated in high school.

The reason that sometimes i am skeptical about ever getting machines of any significant degree of intelligence is that futurists use a definition of "intelligence" that has nothing to do with the definition of intelligence used by ordinary people. When the computer displayed "Would you like to download new updates?" live on the giant screen of the Stanford auditorium so that 200 people could see it and laugh out loud while the elderly physicist was focused on explaining the exciting new findings of the particle accelerator, it was obviously an incredibly stupid moment of an incredibly stupid machine, but futurists would instead point out the number of "logical operations" (ironically abbreviated as FLOPs) that this cheap portable computer can perform in a split second. This will not help build a better machine, just to flop harder (sorry for the pun).

The Accelerating Evolution of Machines

Whenever we look at the rapid progress posted by machines in performing this or that task, it is tempting to say that the machine achieved in a few years what took humans millions of years of evolution to achieve. The argument goes like this: "Yes, it took years to build a machine that recognizes a cat, but how long did it take evolution to create a living being that recognizes cats?"

The truth is that any human-made technology is indirectly using the millions of years of evolution that it took to evolve its creator (Homo Sapiens). No human being, no machine. Therefore it is incorrect to claim that the machine came out of the ENIAC: it came out of millions of years of evolution, just like my nose. The machine that is now so much better than previous models of a few years ago did NOT evolve: WE evolved it (and continue to evolve it).

There is no machine that has created another machine that is superior. WE create a better machine.

We are capable of building machines (and tools in general) because those millions of years of evolution equipped us with some skills (that the machine does NOT have). If humans became extinct tomorrow morning, the evolution of machines would come to an end. Right now this is true of all technologies. If all humans die, all our technologies die with us (until a new form of intelligent life arises from millions of years of evolution and starts rebuilding all those watches, bikes, coffee makers, dishwashers, airplanes and computers). Hence, technically speaking, there has been no evolution of technology.

This is yet another case in which we are applying an attribute invented for one category of things to a different category: the category of living beings evolve, the category of machines does something else, which we call "evolve" by recycling a word that actually has a different meaning. It would be more appropriate to say that a technology "has been evolved" rather than "evolved": computers have been evolved rapidly (by humans) since their invention.

Technologies don't evolve (as of today): we make them evolve.

The day that we have machines that survive without human intervention and build other machines without human intervention, we can apply the word "evolve" to those machines.

As far as i know those machines don't exist yet, which means that there has been zero evolution in machine intelligence so far.

The machine is not intelligent, the engineer who designed it is. That engineer is the product of millions of years of evolution, the machine is a by-product of that engineer's millions of years of evolution.

(See the appendix for a provocative counter-argument: maybe i got it all wrong, and it is technologies that evolve and use us to evolve).

Non-human Intelligence is Already Here

There are already many kinds of intelligence that we cannot match nor truly comprehend. Bats can avoid objects in absolute darkness at impressive speeds and even capture flying insects because their brain is equipped with a high-frequency sonar system. Migratory animals can orient themselves and navigate vast territories without any help from maps. Birds are equipped with a sixth sense for the Earth's magnetic field. Purple martins migrate from Brazil to the USA and back each year. Some animals have the ability to camouflage. The best color vision is in birds, fish, and some insects. Many animals have night vision. Animals can see, sniff and hear things that we cannot, and airports still routinely employ sniffing dogs (not sniffing humans) to detect food, drugs and explosives. And don't underestimate the brain of an insect either: how many people can fly and land upside down on a ceiling?

Howard Hughes' "Sensory Exotica" (1999) and Frans de Waal's "Are We Smart Enough to Know How Smart Animals Are?" (2016) document the amazing skills of the animals that populate our planet.

(The earliest known paintings, in the caves of Lascaux and Chauvet, depict animals, not humans. I suspect those painters marveled at the superhuman powers of animals, not at the superior intelligence of humans).

Virtually all dogs existing today are artificial living beings: they are the result of selective breeding strategies. If you think that your dog is intelligent, then you have "artificial intelligence" right at home.

Ironically, when Deborah Gordon discovered that ant colonies use a packet-switching technique very similar to the one employed by the Internet ("The Regulation of Ant Colony Foraging Activity without Spatial Information", 2012), the media wrote that ants can do what the Internet does when in fact ants have been doing it for about 100 million years: it took human intelligence 200,000 years to figure out the same system of communication devised by ant intelligence.

Summarizing, many animals have powers we don't have. We have arbitrarily decided that any skill possessed by other animals and not by humans is an inferior skill, whereas any skill possessed by humans and not by other animals is a superior skill. This leads me to wonder what will make a skill "superhuman": just the fact that it is possessed by a machine instead of an animal?

And, of course, we already built machines that can do things that are impossible for humans. The clock, invented almost a thousand years ago, does something that no human can do: keeping time. Telescopes and microscopes can see things that humans cannot. We can only see a human-level rendition by those machines, which is equivalent to a higher intelligence explaining something in simpler terms to a lower intelligence. We cannot do what light bulbs do. We cannot touch the groove of a rotating vinyl record and produce the sound of an entire philharmonic orchestra. And, of course, one such appliance is the computer, that can perform calculations much faster than any mathematician could. Even the pre-digital calculators of the 1940s (for example, the ones used to calculate ballistic trajectories) could calculate faster than human brains. In fact, we have always been post-human, coexisting with, and relying on, and being guided by, technology that was capable of super-human feats (and there have always been philosophers debating whether that post-human condition is anti-human or pro-human).

The intelligence of both animals and tools is not called "superhuman" simply because we are used to it. We are not used to robots doing whatever it is that they will do better than us and therefore we call it "superhuman" when in fact we should just call all of these "non-human life"; and maybe "non-human intelligence" depending on your definition of "intelligence".

If a machine ever arises (and proliferates) that is alive and capable of feats that are superhuman, it will just be yet another form of non-human life: not the first one, not the last one. Of course, there are plenty of forms of life that are dangerous to humans, mostly very tiny ones (like viruses and ticks). It comes with the territory. If you want to call it "superhuman", suit yourself.

One gene can make a huge difference in brain structure and function, as the tiny difference between the chimp's DNA and human DNA proves. Gene therapy is already here and that is indeed progressing quickly. Changing the genes of the human DNA may have consequences that are orders of magnitudes bigger than we can imagine. That is one of the reasons why i tend to believe that "superhuman" intelligence, if it comes at all, is more likely to come from synthetic biology than from computers.

There are even qualitative differences in the "intelligences" of a person as the person grows and changes. Psychologists since at least Jean Piaget have studied how the mental life of a child changes dramatically, qualitatively, from one stage in which some tasks are impossible to a new stage in which those tasks become the everyday norm: each new stage represents a "super" intelligence from the viewpoint of the previous stage. There is an age at which the child conceives little more than herself and her parents. That child's brain just cannot conceive that there are other people and that people live on a planet and that the planet contains animals, trees, seas, mountains, etc; that you have to study and work; not to mention the mind-boggling affairs of sex and where children come from; and that some day you will die. All of this emerges later in life, each stage unlocking a new dimension of understanding. (And i wonder if there is an end to this process: if we lived to be 200 years old in good health, what would be our understanding?) My intelligence is "super" compared to the intelligence that i had as a little child.

At the same time try learning languages or any other skills at the speed that children learn them. Children can do things with their minds that adults cannot do anymore: sometimes you feel that you cannot understand what their minds do, that they are little monsters. Children are superhuman too, as Alison Gopnik argues in "The Philosophical Baby" (2009). One wonders what we could achieve if we remained children all our lives (more about this later).

The Consciousness of Super-human Intelligence

(Warning: this chapter and the next one are boring philosophical speculation).

Given that non-human intelligence exists all around us, what would make a particular non-human intelligence also "superhuman"? I haven't seen a definition of "superhuman" (as opposed to simply "non-human").

However, there is at least one feature that i would expect to find in a superhuman intelligence: consciousness. I think, i feel, sometimes i suffer and sometimes i rejoice. If i have consciousness, an intelligence that is superior to mine should have it too.

We know that human brains are conscious, but we don't really know why and how. We don't really know what makes us conscious, how the electrochemical processes inside our brain yield feelings and emotions. (My book "Thinking about Thought" is a survey of the most influential viewpoints on consciousness). An electronic replica of your brain might or might not be conscious, and might or might not be "you". We don't really know how to build conscious beings, and not even how to find out if something is conscious. If one of the machines that we are building turns out to develop its own consciousness, it will be an amazing stroke of luck.

However, i doubt that i would call "superhuman" something that is less conscious than me, no matter how fast it is at calculating the 100th million digit of the square root of 2, how good it is at recognizing cats and how good it is at playing go/weichi.

However, you might object, a super-human intelligence will not need to be conscious. You might object that feelings and emotions are a sign of weakness, not of strength. Consciousness makes us cry. Feelings cause us to make mistakes that we later regret, and that sometimes hurt us or hurt others. Maybe a being that is more intelligent than us and does not feel anything is actually the secret to outperforming human intelligence.

In fact, "consciousness" for an information-processing machine could be something altogether different from the consciousness of an energy-processing being like us. Our qualia (conscious feelings) measure energy levels: light, sound, etc. If information-processing machines ever develop qualia, it would make sense that those qualia be about information levels; not qualia related to physical life, but qualia related to "virtual" life in the universe of information.

It is not even clear whether superhuman intelligence requires human intelligence first: can human-level intelligence be skipped on the way to superhuman intelligence? Do machines need to be as smart as us before becoming smarter than us or can they find a short cut to superhuman intelligence?

We cannot answer this question looking at biological intelligence because the progress of machine intelligence is happening in a completely different way from the way that biological intelligence evolved. The way Nature works is simple: new species don't need to climb the ladder of intelligence: they start out at a given level of intelligence, bypassing all the lower ones. For example, humans have never been so little intelligent as bacteria. The way Artificial Intelligence works is different: it tweaks software programs, making them more and more intelligent, and these software programs can run on any computer that is powerful enough. A.I. is about the progress of software (that can run on any hardware), Nature is about the progress of hardware (a hardware that also includes a brain that, in turn, somehow includes a software called "mind").

The Intelligence of Super-human Intelligence

What does it take for machine intelligence to reach the point of human-level intelligence?

One is tempted to answer: just build an electronic replica of a human brain. If we replaced each and every neuron in your brain with an electronic chip, i am not sure that you would still be "you" but your brain should still yield a form of human intelligence, wouldn't it?

Unfortunately, we are pretty far from implementing that full replica of your brain (note that i keep saying "your" and not "mine"). It is a bit discouraging that the smallest known brain, the brain of the roundworm (300 neurons connected by a few thousand synapses) is still smarter than the smartest neural network ever built.

If you think that this hypothetical electronic replica of your brain would not be as smart as you, then you are implying that the very "stuff" of which the brain is made is important in itself; but then machine intelligence is impossible because machines are not made of that "stuff".

And, again, can a machine reach human-level intelligence without consciousness? Is consciousness required in order to be as smart as Einstein? Could a machine be as smart as Einstein without having any feelings and emotions?

Does "machine intelligence" require all of you, including the mysterious inscrutable silent existence that populates your skull, the vast unexplored land of unspoken thoughts and feelings that constitutes "you"? What i hear when i listen to you is just a tiny fraction of what you are thinking and feeling. What i see when i watch you is just a tiny fraction of what you thought of doing, dreamed of doing, and plan doing. When i hear a robot talk, that is the one and only thing that it "wants" to say. When i see it move, that is the one and only thing that it wants to do.

Intelligent Behavior in Structured Environments

When you need to catch a bus in some underdeveloped countries, you don't know at what time it will arrive nor how much you will be charged for the ticket. In fact you don't even know how it will look like (it could be a generic truck or a minivan) and where it will stop. Once on board, you tell the driver where you want to get off and hope that he will remember. If she is in a good mood, she might even take a little detour to drop you right in front of your hotel. On the other hand, when you take a bus in a developed country, there is an official bus stop (the bus won't stop if you are 20 meters before or after it), the bus is clearly recognizable and marked with the destination and won't take any detour for any reason, the driver is not allowed to chat with the passengers (sometimes she is physically enclosed in a glass cage), the ticket must be bought with exact change at a ticket vending machine (and sometimes validated inside at another machine). There is a door to be used to exit, and you know when to exit because the name of the bus stop is displayed on a LED screen. On many long-distance trains and buses you also get an assigned seat (you can't just sit anywhere).

It is easy to build a robot that can ride a bus in a developed country, much more difficult to build a robot that can ride a bus in an underdeveloped country. What makes it easy or difficult is the environment in which it has to operate: the more structured the environment, the easier for the robot. A structured environment requires less "thinking": just follow the rules and you'll make it. However, what really "makes it" is not you: it's you plus the structured environment. That's the key difference: operating in a chaotic, unpredictable situation is not the same thing as operating in a highly structured environment. The environment makes a huge difference. It is easy to build a machine that has to operate in a highly structured environment, just like it is easy for a bullet train to ride at 300 km/hour on rails.

We structure the chaos of nature because it makes it easier to survive and thrive in it. Humans have been spectacularly successful at structuring their environment so that it obeys simple, predictable rules. This way we don't need to "think" too much: the structured environment will take us where we want to go. We know that we can find food at the supermarket and a train at the train station. In other words, the environment makes us a little more stupid but allows anybody to achieve tasks that would otherwise be difficult and dangerous, i.e. that would require a lot of intelligence. When the system fails us, we get upset because now we have to think, we have to find a solution to an unstructured problem.

If you are in Paris and the metro is on strike and it is impossible to get a taxi, how to do you get to your appointment in time? Believe it or not, most Parisians manage. Most tourists from the USA don't. If there is no traffic light and cars don't stop for pedestrians and traffic is absolutely horrible, how do you cross a wide boulevard? Believe it or not, Iranians do it all the time. Needless to say, most Western tourists spend hours trying to figure it out.

It is certainly very impressive how well humans structure a universe that is chaotic. The more we structure it, the easier for extremely dumb people and machines to survive and thrive in it.

The claims of the robotic industry are often related to structured environments, not to their robots. It is relatively easy to build an autonomous car that rides on a highway with clearly marked lanes, clearly marked exits, ordered traffic, and maps that detail everything that is going to happen. It is much more difficult (orders of magnitude more difficult) to build an autonomous car that can drive through Tehran or Lagos (this is a compliment to Iranian and Nigerian drivers, not an insult). Whoever claims that a computer is driving a car is distorting the facts: it is not the computer that is driving the car but the environment that has been structured so that any inexperienced and not particularly intelligent driver, and even a computer, can drive a car. Today's computer cannot drive a car in the traffic of Lagos or Tehran. It will if and when the streets of Lagos and Tehran become as well structured as the streets of California, if and when Iranian and Nigerian drivers are forced to obey strict traffic rules. Saying that the on-board computer is steering the driverless car is like saying that the locomotive knows in which direction to take the train: the locomotive is simply constrained by the rails to take the correct direction.

In order for self-driving cars to use our streets, we will need to retrofit roads with devices that tell the car what to do at every point in time. It is not intelligence but old-fashioned infrastructure that will allow very dumb self-driving cars to drive safely; in other words we will need the equivalent of the highly-structured system of rails and controllers that will, eventually, allow fast, safe and accurate trains to run.

I recently had to exchange the equivalent of $3.00 in a local currency while leaving a Western country at its capital's airport. The procedure was silly beyond belief. I had to produce passport, boarding pass and receipt of previous money exchanges before getting my money, a lengthy operation for just three dollars. On the contrary at the border between Haiti and Dominican Republic, a wildly chaotic place with taxi drivers, fruit vendors and police officers yelling at each other and at everybody passing by, there was a mob of money changers chasing the travelers. I had to guess which ones were honest money changers rather than scammers, and then bargain the exchange rate, and then make sure that the money was good while all the time protecting my wallet from pickpockets. It wouldn't be difficult to build a robot that can exchange money at the airport of a Western capital, but orders of magnitude more difficult to build one that can exchange money while walking from the immigration post of Haiti to the immigration post of the Dominican Republic.

The more structured the environment, the easier it is to build a machine that operates in it. What really "does it" is not the machine: it's the structured environment. What has made so many machines possible is not a better A.I. technology, but simply better structured environments. It's the rules and regulations that allow the machine to operate.

You can't call an automatic phone system and just explain your problem. You have to press 1 for English, 1 for customer support, 3 for your location, 2 for your kind of problem and 4 and 7 and so forth. What allows the machine to perform its job, and to replace the human operator, is that you (the human being) have removed the human aspect from the interaction and behave like a machine in a mechanical world. It is not the machine that behaves like a human being in a human world.

The fundamental thing that a self-driving car must be able to do is, of course, to stop at a gas station when it runs out of gasoline. Can these cars autonomously enter a gas station, stop in front of a pump, slide a credit card in the payment slot, pull out the hose and pour gasoline in the tank? Of course, not. What needs to be done is to create the appropriate structured environment for the driverless car (or, better, for some sensors on board the car) so that the car will NOT need to behave like an intelligent being. The gas station, the gas pump and the payment used by the driverless car will look very different from the one used so far by human drivers.

Incidentally, most of those rules and regulations that create a highly structured environment (favorable to automata) were originally introduced in order to reduce costs. Employing machines has been the next logical step in cost reduction. The machine is one step in an ongoing process of cost reduction and productivity increase. The goal was not to create superhuman intelligence, just to increase profits.

Think of your favorite sandwich chain. You know exactly what kind of questions they will ask you. There is a well-structured process by which your sandwich will be made. The moment robots become cheap enough, they will certainly take over the jobs of the kids who prepare your sandwich today. It is not a matter of "intelligence" (the intelligence of today's robots is already more than enough) but of cost: today a teenager is cheaper than a robot. The whole point of structuring the sandwich-making process was to allow inexperienced and unskilled workers (read: underpaid) to perform the task once reserved for skilled experienced chefs.

The more unstructured the environment is, the more unlikely that a machine will replace the human. Unfortunately, one very unstructured environment is that of health care. Medical records are kept on physical files, and doctor's notes are notoriously impossible to read. There is very little that a machine can do in that environment. The way to introduce "intelligent" machines in that environment is, first of all, to structure all that information. When it is "digitized" and stored in databases, it means that it has been structured. At that point any human being, even with little or no knowledge of medical practice, can do something intelligent in that environment. And even a machine can.

The truth is that we do not automate jobs as they are. First, we dehumanize the job, turning it into a mechanical sequence of steps. Then we use a machine to automate what is left of that job. For example, my friend Steve Kaufman, a pediatrician all his life, realized that his skills were less and less necessary: a nurse practitioner can fill all the forms and click on all the computer buttons that are required when seeing a patient; the doctor, who is increasingly required to type on a keyboard, may not even make eye contact with the patient. This has the beneficial effect of reducing the number of days that the average patient spends at a hospital, but it erases the kind of bonding between doctor and patience that was common in the "unstructured" world. When the last vestiges of humanity will have been removed from the job of the doctor, it will be relatively easy to automate the doctor's job. But that is not what Steve was doing. As Steve pointed out to me, if you don't bond with an asthmatic patient, you may never realize that he is suicidal: you will cure his asthma, but he will commit suicide; and the machine will still archive the case as a success.

Structured environments are also relying on ever stricter rules. My favorite example is the boarding procedure at an airport, where we are treated like cattle from check-in to the gate, with a brief interval during which we are treated like a walking credit card that airport shops desperately try to get. Other than the credit card thing, we are basically building the kind of hyper-bureaucratic state pioneered by the Soviet Union.

There is a fundamental paradox underlying the ongoing structuring of society. What is profoundly human (and actually shared by all forms of life) is the vagueness of language and behavior. What humans (and animals) can do relatively well, and do on a daily basis, and today's machines are not good at, is to deal with ambiguity. Unfortunately, ambiguity is responsible for a lot of the miscommunication and chaos that complicate our life. Rules and regulations are useful because they remove ambiguity from society, and therefore simplify our life. As a side-effect, though, the more we structure human behavior by removing ambiguity, the more replicable it becomes. We become machines; machines that demand a high salary and all sorts of rights. It is a no-brainer for businesses to replace such expensive machines with cheaper ones that don't demand any right.

Increasingly structured environments, routines and practices will eventually enable the automation of "cognitive" skills too. I am writing while watching the indecent spectacle of the political campaigns for a presidential election in the USA. Political debates are becoming more and more structured, with a format agreed beforehand and a moderator that enforces it, and a restriction on the kind of questions that can be asked, and candidates who basically memorize press releases worded by their campaign staff. It is not difficult to imagine that sooner or later someone will build a piece of software that can credibly replace a politician in a political debate; but that feat will owe more to the lack of real debate in these political debates than to greater rhetorical skills on the part of the machine. On the other hand that software will be incapable of participating in a passionate conversation about a World Cup game with a group of rowdy and drunk soccer fans.

It is the increasingly structured environment that is enabling and will enable the explosion of robotics and automated services. Most of the robots and phone-based services coming to the market now rely on relatively old technology. What has made them feasible and practical is that they can now operate in highly structured environments.

Think of yourself. You are now identified by numbers in so many different contexts: your passport number, your social security number, your street address, your telephone number, your insurance policy number, your bank account number, your credit card number, your driver license number, your car's plate number, your utility bill account number... It is a rarity when someone tries to identify me based on non-numeric features. And increasingly we depend on passwords to access our own information. The more we reduce the individual to a digital file, the easier it gets to build "intelligent assistants" for that file... sorry, i meant "for that person".

In a sense, humans are trying to build machines that think like humans while machines are already building humans who think like machines.

Intermezzo: Will Intelligent Machines Return to Chaotic Environments?

Structuring the environment really consists of two parallel processes. On the one hand, it means removing the chaotic and unpredictable (and often intractable) behavior of natural environments. On the other hand, it also means removing the chaotic and unpredictable (and often intractable) behavior of human beings. The purpose of all the rules and regulations that come with a structured environment is to replace you (a messy human intelligence) with an avatar that is like you (in fact it shares your body and brain) without the quirkiness of human intelligence. That avatar lives in a highly-structured virtual world that mimics the natural world without all the quirkiness of the (wildly unstructured) natural world.

My thesis is that machines are not becoming particularly more intelligent, but, instead, it is humans who are structuring the environment and regulating behavior so that humans become more like machines and therefore machines can replace humans.

But what happens if machines become truly "intelligent"? If "intelligent" means that machines will become what humans still are today, before society turns them into rule-obeying machines, then, ironically, machines may acquire all the "baggage" that intelligent biological beings carry, i.e. the unpredictable, chaotic, anarchic behavior that any living being exhibits, i.e. precisely what the structured environment and rules and regulations aim at suppressing.

It would be ironic if creating intelligent machines would turn machines into (messy) humans at the same time that we are turning humans into (disciplined) machines.

Another Intermezzo: Disorder is Evolution, Order is Stagnation

Equilibrium is not a normal state in the universe. The universe is a vast collection of "open" systems that trade energy, matter and information with each other. Many systems thrive in a state far from equilibrium, the so-called "edge of chaos". Living beings are an example: living beings trade energy, matter and information with their ecosystem. You are constantly living at the "edge of chaos". You will reach a state of equilibrium when you die. We can view intelligent systems as systems that are particularly complex.

Ilya Prigogine, Stuart Kauffman and many others have shown an interesting property of these systems. Complex systems (technically speaking, "nonlinear" systems) that are pushed out of equilibrium by perturbations reach a point where they can either disintegrate in total chaos or spontaneously reorganize themselves at a higher level of complexity. The outcome is unpredictable, and irreversible.

A rigid society, in which rules and regulations enforce some behavior and prohibit some other behavior, leaving little to the imagination, is not a complex system. It is very predictable in what happens to you if you break those rules: you go to jail, or you get fired.

"Noise" (perturbations) is important in self-organizing systems such as the human society because it allows such systems to evolve. Under the right circumstances a self-organizing system disturbed by noise will self-organize at a higher level, in some cases a level that is profoundly different from the original one. The more we remove "noise" and unpredictability from human society, the less likely that human society will evolve at all, let alone towards higher levels of organization.

Intelligent machines might rediscover this law of non-equilibrium thermodynamics after humans have forgotten it.

Human Obsolescence

Both computer experts and ordinary people fear that we (humans) may become obsolete because machines will soon take our place.

Jack Good wrote in "Speculations Concerning the First Ultraintelligent Machine" (1965): "the first ultraintelligent machine is the last invention that man need ever make". Hans Moravec in "Mind Children" (1988): "Robots will eventually succeed us: humans clearly face extinction". A 2000 article by Bill Joy was titled "The Future doesn't Need us". Etcetera. Actually, this idea has been repeated often since the invention of (among other things) the typewriter and the assembly line.

When we say that "robots will succeed us" or "The future doesn't need us", we really need to define "us". Assembly lines, typewriters, computers, search engines, steam engines, printing presses and whatever comes next have replaced jobs that have to do with material life. I could simply say that they have replaced "jobs". They have not replaced "people". They replaced their jobs. Therefore what went obsolete has been jobs, not people, and what is becoming obsolete is jobs, not people. Humans are biological organisms who (and not "that") write novels, compose music, make films, play soccer, root for Tour de France bicyclists, discover scientific theories, argue about politics, hike on mountains and dine at fancy restaurants. Which of these activities is becoming obsolete because machines are doing them better?

Machines are certainly good at processing big data at lightning speed. Fine. We are rapidly becoming obsolete at doing that. In fact, we've never done that. Very few humans spent their time analyzing big data. The vast majority of people are perfectly content with small data: the price of gasoline, the name of the president, the standings in the soccer league, the change in my pocket, the amount of my electricity bill, my address, etc. Humans have mostly been annoyed by big data. That was, in fact, a motivation to invent a machine that would take care of big data. The motivation to invent a machine that rides the Tour de France is minimal because we actually enjoy watching (human) riders sweat on those steep mountain roads, and many of us enjoy emulating them on the hills behind our home. Big data? Soon we will have a generation that cannot even do arithmetic.

What is becoming obsolete is not "us" but our current jobs. That has been the case since the invention of the first farm (that made the prehistoric gatherers obsolete) and, in fact, since the invention of the wheel (the cart made many porters obsolete), and jobs certainly disappeared when Gutenberg started printing books with the printing press, the precursor of the assembly line.

Since then, humans have used wheels to travel the world and the printing press to discuss philosophy.

In Defense of Progress: Augmented Intelligence

Enough of bashing computers. The computer might be the only major appliance invented since television, but it is qualitatively different than all the previous appliances. What can the dishwasher do other than wash dishes? The computer, instead, can do a lot of things, from delivering mail to displaying pictures. A computer is many machines in one. That was, in fact, the whole point of the Universal Turing Machine: a universal problem solver. Little did he know that its applications would range from phone conversations to social media.

The secret is the software:

In fact, there has been little progress in the physical world but a lot in the virtual world created by computers. Just witness the explosion of online services in the 1990s and of smartphone applications since 2007.

Perhaps even more importantly, the law of entropy does not apply to software: everything in this universe is bound to decay and die because of the second law of Thermodynamics (that entropy can never decrease). That does not apply to software. Software will never decay. Software can create worlds in which the second law of Thermodynamics does not apply: software never ages, never decays, never dies. (Unfortunately, software needs hardware to run, and that hardware does decay).

The catch is that software does not have a body and therefore cannot do anything unless it is attached to a machine. Software cannot cook and cannot start a car unless we drop it inside a computer and attach the computer to the appropriate machine. Software cannot even give answers without a printer, a screen or a speaker.

Disembodied software is like disembodied thought: it is an abstraction that doesn't actually exist.

Software has to be incorporated into a processor in order to truly exist (to "run"). In turn the processor, that ultimately only does binary algebra, has to be attached to another machine in order to perform an action, whether cooking an omelette or starting a car.

De facto, we attach a universal problem solver to a specific problem solver. However, there is a way to maximize the usefulness of a universal problem solver: attach it to another universal problem solver, the human mind.

One could argue that, so far, Artificial Intelligence has failed to deliver, but "Augmented Intelligence" has been successful beyond the hopes of its founding fathers. In the 1960s in Silicon Valley there were two schools of thought. One, usually associated with John McCarthy's Stanford Artificial Intelligence Lab (SAIL), claimed that machines would soon replace humans. The other one, mainly associated with Doug Engelbart at the nearby Stanford Research Institute (now SRI Intl), argued that machines would "augment" human intelligence rather than replace it. Engelbart's school went on to invent the graphic user interface, the personal computer, the Internet, and virtual personal assistants like Siri; all things that "augmented" human intelligence. This program did not necessarily increase human intelligence and it did not create a non-human intelligence: the combination of human intelligence and these devices can achieve "more" than human intelligence can alone.

The search engine is a good example of "amazing" augmented intelligence and "disappointing" artificial intelligence. It must be terribly difficult for search engines to keep up with the exponential growth of user-provided content. The ranking algorithm has to become exponentially smarter in order for the search engine to keep providing relevant answers. It's something that the user doesn't see (unlike, say, a new button on the microwave oven), but it's something that is vital to making sure that the World-wide Web does not become unsearchable, i.e. a "World-wide Mess".

General-purpose Intelligence

Before analyzing what it will take (and how long it will take) to get machine intelligence, we need to define what we are talking about.

A man, wearing a suit and tie, walks out from a revolving hotel door dragging his rolling suitcase. Later, another man, wearing a shabby uniform and gloves, walks out of the side door dragging a garbage can. It is obvious even to the dumbest human being that one is a guest of the hotel and the other one is a janitor. Do we require from a machine this simple kind of understanding ordinary situations in order for it to qualify as "intelligent"? Or is it irrelevant, just like matching the nightingale's song is irrelevant in order to solve differential equations? If we require that kind of understanding, we push machine intelligence dramatically forward into the future: just figuring out that one is a suit and tie and one is a uniform is not trivial at all for a machine. It takes an enormous computational effort to achieve just this one task. There are millions of situations like this one that we recognize in a split second.

Let us continue our thought experiment. Now we are in an underdeveloped country and the janitor is dragging an old broken suitcase full of garbage. He has turned an old suitcase into his garbage can. Seeing such a scene, we would probably just smile at the man's ingenuity; but imagine how hard it is for a machine to realize what is going on. Even if the machine is capable of telling that someone dragging a suitcase is a hotel guest, the machine now has to understand that a broken suitcase carried by a person in a janitor's uniform does not qualify as a suitcase.

There are millions of variants on each of those millions of situations that we effortlessly understand, but that are increasingly trickier for a machine.

The way that today's A.I. scientists would go about it is to create one specific software program for each of the millions of situations, and then millions of their variants. Given enough engineers, time and processors, this is feasible. Whenever a critic like me asks "but can your machine do this too?", today's A.I. scientists rush out to create a new program that can do it. "But can your machine also do this other thing?" The A.I. scientists rush out to create another program. And so forth.

Given enough engineers, time and processors, it is indeed possible to create a million machines that can do everything we naturally do.

After all, the Web plus a search engine can answer any question: someone, sooner or later, will post the answer on the Web, and the search engine will find it. Billions of Web users are providing all the answers to all the possible questions. The search engine is not particularly intelligent in any field but can find the answer to questions in all fields.

I doubt that this is the way in which my mind works (or any animal's mind works), but, yes, those millions of software programs will be "functionally" equivalent to my mind. In fact, they will be better than my mind because they will be able to recognize all the situations that all the people in the world recognize, not just the ones that i recognize, just like the Web will eventually contain the answers to all questions that all humans can answer, not only the answers that i know.

This is exactly what "brute-force A.I." is doing today: creating a specific software program for each intelligent task that humans perform. The method is different, but the rationale is reminiscent of Marvin Minsky's "The Society of Mind" (1985) that viewed an artificial general intelligence as a society of specialized agents.

Luckily, the effect on the economy will be to create millions of jobs because those millions of machines will need to be designed, tested, stored, marketed, sold, and, last but not least, repaired.

The Proliferation of Appliances, Intelligent and not Intelligent

If we structure the world appropriately, it will be easy to build machines that can board planes, exchange money, take a bus, drive a car, cross a street and so on. Automated services have existed since at least the invention of the waterwheel. We even have machines that dispense money (ATMs), machines that wash clothes (washing machines), machines that control the temperature of a room (thermostats), and machines that control the speed of a car (cruise controls).

When we design robots, we are simply building more appliances. In the near future we might witness a multiplication of appliances, disguised and marketed as "robots" simply because the word "robot" is becoming fashionable: iRobot's vacuuming robot Roomba, Moley Robotics' robotic chef in Britain that, installed on top of your stove, cooks dinner for you; the robotic waiters of the Robot Restaurant in Harbin (northeastern China); Infinium Robotics' drone waiters, that deliver meals flying over the heads of the customers, in Singapore; MIT's robotic bartender; UC Berkeley's robot that folds towels; etc.

The ATM is more precise than a bank clerk (and works much longer hours) but we don't think of it as "intelligent". Ditto for the washing machine that is capable of all sorts of washing techniques. That's because they were introduced at a time when it was not popular to market them as Artificial Intelligence. If the washing machine was invented today, it would certainly be presented as the latest achievement in robotics.

Enthusiastic fans of automation predict that "soon" (how soon?) everything that humans do will be done by machines; but they rarely explain what is the point of making machines for everything we do. Do we really want machines that fall asleep or urinate? There are very human functions that people don't normally associate with "intelligence". They just happen to be things that human bodies do. We swing arms when we walk, but we don't consider "swinging arms while walking" a necessary feature of intelligent beings. The moment we attempt to design an "intelligent" machine (or collection of machines) that can mimic the entire repertory of our "intelligent" functions we run into the enumeration problem: which function qualifies as "intelligent"? Typical human activities include: forgetting where we left the mobile phone, eating fast food, watching stand-up comedy, catching a flue when attacked by viruses and, yes, frequently, urinating.

We instinctively envision a hierarchy of tasks, from "not intelligent at all" to "very intelligent", and we assume that the latter are the ones that make the difference. However, that ranking is not very objective: why a washing machine is not intelligent given that relatively few humans can wash clothes, whereas a cat-recognizing program is (given that virtually every human, no matter how dumb, can recognize a cat, and so can countless animals). Statistically, it would seem that washing clothes should be more special than cat-recognizing programs.

The current excitement about machines is due to the fact that (it is claimed) they are beginning to perform tasks that were exclusive to human beings. This is actually a very weak claim: the first washing machine was capable of performing a task that had been exclusive to human beings until the day before. Implicit in these claims is the idea that there is something that makes some tasks qualitatively more "special" than washing clothes, but it is difficult to articulate what this "special" quality would be. What is truly unique/special about human intelligence? Each machine performs for us a task that we used to do manually. Which tasks are so "special" that they deserve to be called "intelligent" is far from agreed upon.

And, finally, machines that resemble human beings (that smile, cry, walk and even say a few words) have existed for a long time and they are usually sold in toy stores and aimed at children. We can certainly create more sophisticated toys, like toys that recognize cats, but the claim that these toys will have anything to do with human intelligence needs some explaining.

Intermezzo: The Resurrection of the Dead

For as long as we have been making tools, technology has been the language of the dead speaking to the living. Maybe we are fascinated by intelligent machines because we would finally have technology that is alive. As more and more technology has invaded our private and public lives, it has become a bit discomforting to be talking so often to the dead.

Demystifying the Turing Test

The Turing Test is the best known test to determine whether a machine has become as intelligent as humans: a person asks questions until it can tell whether the answers are coming from a human or a machine (that must not be visible, of course). If that person cannot reach a conclusion (or reaches the wrong conclusion), the machine has passed the test. Any apprentice philosopher can tell you that it all depends on the questions that are being asked. If you ask the questions that make us human, all computer programs fail the Turing Test, and they fail in awkward manners.

Linguists like to talk about the difficulty of understanding ambiguous sentences such as "Prostitutes appeal to Pope" and "Iraqi head seeks arms". But the job of a machine gets even more difficult when common sense is involved. In the sentence "Carl, who died last year, was a great scientist, and his son Dale has fond memories, and he now takes care of the center" it is pretty clear to whom the "he" refers, because one of the two men is dead and therefore he cannot take care of the center (or of anything else). This is not obvious to a machine that doesn't know what "dying" implies.

Ask the machine: "The doll will not fit in the box because it is too big: which one is too big, the doll or the box?" If you ask questions like this one, the human being will get them right almost 100% of the time, but the machine will only get them right 50% of the time because it will simply be guessing (like flipping a coin). Ask just two sentences like this one, and, most likely, you will know whether you are talking to a machine or to a human being. The machine has no common sense: it doesn't know that, in order to fit inside a box, an object has to be smaller than the box. This is the essence of the Winograd Schema Challenge devised by Hector Levesque, at the University of Toronto in 2011.

Common Sense

In November 1958, at the Symposium on Mechanization of Thought Processes in England, the always prescient John McCarthy delivered a lecture titled "Programs with Common Sense", that became one of the most influential papers in A.I. McCarthy understood that a machine with no common sense is what we normally call "an idiot". It can certainly do one thing very well, but it cannot be trusted to do it alone, and it certainly cannot be trusted doing anything else.

What we say is not what we mean. If I ask you to cook dinner using whatever high-protein food you can find in a kitchen cabinet, that does not mean that you should cook the spider crawling on its walls, nor the chick that your children have adopted as a pet, nor (gasp) the toddler who is hiding in it for fun.

How do we decide when is the best time to take a picture at an event? A machine can take thousands of pictures, one per second, and maybe even more, but we only take 2 or 3 because those are the meaningful events.

Surveillance cameras and cameras on drones can store millions of hours of videos. They can recognize the make and model of a car, and even read its plate number, but they can't realize that a child is drowning in a swimming pool or that a thief is breaking into a car.

In April 2016 in England a group of children spontaneously formed a human arrow on the ground to direct a police helicopter towards the fleeing suspects of a crime. Nobody taught the children to do that. What the children guessed (in a few seconds) is a fairly long list of "common sense" knowledge: there has been a crime and we need to capture the criminals; the criminals are running away to avoid capture; the helicopter in the sky is the police looking for the criminals; the police force is the entity in charge of catching criminals; it is good that you help the police if you have seen the criminals flee; it is bad if the criminals escape; the helicopter cannot hear you but can see you if you all group together; the arrow is a universal symbol to mark a direction; helicopters fly faster than humans can run; etc. That's what intelligence does when it has common sense.

Around the same time in 2016 Wei Zexi, a 21-year-old student from Xidian University in China's Shaanxi province, who was undergoing treatment for a rare form of cancer, found an ad on Baidu (China's search engine) publicizing a treatment offered by the Beijing Armed Police Corps No 2 Hospital. The "doctor" turned out to be bogus and the treatment killed the boy. The Chinese media demonized Baidu (and, hopefully, the military hospital!), but this was not a case of Baidu being evil: it was the case of yet another algorithm that had no common sense, just like the Google algorithm that in 2015 thought two African-Americans were gorillas, just like the Microsoft algorithm that in 2016 posted racist and sexist messages on Twitter. This is what intelligence does when it has no common sense.

To make things worse, i found the news of Wei Zexi's death on a website that itself displayed some silly ads. Two of these ads were almost porno in nature (titled "30 Celebs Who Don't Wear Underwear" and "Most Embarrassing Cheerleader Moments"). These ads were posted next to the article describing the tragic death of Wei Zexi: the "intelligent" software that assigns ads to web pages has no common sense, i.e. it cannot understand that it is really disgusting to post such sex-related ads in a page devoted to someone's death. (No, the ads were not customized for me: i was using an Internet-café terminal).

When computers became powerful enough, some A.I. scientists embarked upon ambitious attempts to replicate the "common sense" that we humans seem to master so easily as we grow up. The most famous project was Doug Lenat's Cyc (1984), which is still going on. In 1999 Marvin Minsky's pupil Catherine Havasi at the MIT launched Open Mind Common Sense that has been collecting "common sense" provided by thousands of volunteers. DBpedia, started at the Free University of Berlin in 2007, collects knowledge from Wikipedia articles. The goal of these systems is to create a vast catalog of the knowledge that ordinary people have: plants, animals, places, history, celebrities, objects, ideas, etc. For each one we intuitively know what to do: you are supposed to be scared of a tiger, but not of a cat, despite the similarities; umbrellas make sense when it rains or at the beach; clothes are for wearing them; food is for eating; etc. More recently, the very companies that are investing in deep learning have realized that you can't do without common sense. Hence, Microsoft started Satori in 2010 and Google revealed its Knowledge Graph in 2012. By then Knowledge Graph already contained knowledge about 570 million objects via more than 18 billion relationships between objects (Google did not disclose when the project had started). These projects marked a rediscovery of the old program of "knowledge representation" (based on mathematical logic) that was downplayed too much after the boom in deep learning. Knowledge Graph is a "semantic network", a kind of knowledge representation that was very popular in the 1970s. Google's natural-language processing team, led by Fernando Pereira, is integrating Google's famous deep-learning technology (the "AlphaGo" kind of technology) with linguistic knowledge that is the result of eight years of work by professional linguists.

It is incorrect to say that deep learning is a technique for learning to do what we do. If i do something that has never been done before, deep learning cannot learn how to do it: it needs thousands if not millions of samples in order to learn how to do it. If it is the first time that it has been done, by definition, deep learning cannot learn it: there is only one case. Deep learning is a technique for learning something that humans DID in the past.

Now let's imagine a scenario in which neural networks have learned everything that humans ever did. What happens next? The short answer is: nothing. These neural networks are incapable of doing anything that they were not trained to do, so this is the end of progress.

Training a neural network to do something that has never been done before is possible (for example, you can just introduce some random redistribution of what it has learned), but then the neural network has to understand that the result of the novel action is interesting, which requires an immense knowledge of the real world. If I perform a number of random actions, most of them will be useless, wasteful of time and energy, but maybe one or two will turn out to be useful. We often stumble into interesting actions by accident and realize that we can use those accidental actions for doing something very important. I was looking for a way to water my garden without having to physically walk there, and one day i realized that an old broken hose had so many holes in it that would work really well to water the fruit trees. Minutes ago, i accidentally pressed the wrong key on my Android tablet and discovered a feature that I didn't know existed. It is actually a useful feature.

In order to understand which novel action is useful, one needs a list of all the things that can possibly be useful to a human being. It is trivial for us to understand what can be useful to human life. It is not trivial for a machine, and certainly not trivial at all for a neural network trained to learn from us.

See for example Alexander Tuzhilin's paper "Usefulness, Novelty, and Integration of Interestingness Measures" (Columbia

University, 2002) and Iaakov Exman's paper "Interestingness a Unifying Paradigm Bipolar Function Composition" (Israel, 2009).

The importance of common sense in daily activities is intuitive. We get angry whenever someone does something without "thinking". It is not enough to recognize that a car is a car and a tree is a tree. It is also important to understand that cars move and trees don't, that cars get into accidents and some trees bear edible fruits, etc. Deep learning is great for recognizing that a car is a car and a tree is a tree, but it struggles to go beyond recognition. So there is already a big limitation.

A second problem with deep-learning systems is that you need a very large dataset to train them. We humans learn a new game just from listening to a friend's description and from watching friends play it a couple of times. Deep learning requires thousands if not millions of cases before it can play decently.

Big data are used to train the neural networks of deep learning systems, but "big data" is not what we use to train humans. We do exactly the opposite. Children's behavior is "trained" by two parents and maybe a nanny, not by videos found on the Internet. Their education is "trained" by carefully selected teachers who had to get a degree in education. We train workers using the rare experts in the craft, not a random set of workers. We train scientists using a handful of great scientists, not a random set of students.

I am typing these words in 2016 while Egypt and other countries are searching the Mediterranean Sea for an airplane that went missing. In 2014 a Malaysia Airlines airplane en route from Kuala Lumpur to Beijing mysteriously disappeared over the Indian Ocean. Deep-learning neural networks can be trained to play go/weichi because there are thousands of well documented games played by human masters, but the same networks cannot be trained to scour the ocean for debris of a missing airplane: we don't have thousands of pictures of debris of missing airplanes. They can have arbitrary shapes, float in arbitrary ways, be partially underwater, etc. Humans can easily identify pieces of an airplane even if they have only seen 10 or 20 airplanes in their life, and never seen the debris of an aircrash; neural networks can only do it if we show them thousands of examples.

A third problem related to machines with no common sense is their inability to recognize an "obvious" mistake. Several studies have shown that, in some circumstances, deep-learning neural networks are better than humans at recognizing objects; but, when the neural network makes a mistake, you can tell that it has no common sense: it is usually a mistake that makes us laugh, i.e. a mistake that no idiot would make. You train a neural network using a large set of cat photos. Deep learning is a technique that provides a way to structure the neural network in an optimal way. Once the neural network has learned to recognize a cat, it is supposed to recognize any cat photo that it hasn't seen before. But deep neural networks are not perfect: there is always at least one case (a "blind spot") in which the neural network fails and takes the cat for something else. That "blind spot" tells a lot about the importance of common sense. In 2013 a joint research by Google, New York University and UC Berkeley showed that tiny perturbations (invisible to humans) can completely alter the way a neural network classifies the image. The paper written by Christian Szegedy and others was ironically titled "Intriguing Properties Of Neural Networks". Intriguing indeed, because no human would make those mistakes. In fact, no human would notice anything wrong with the "perturbed" images. This is not just a theoretical discussion. If a self-driving car that uses a deep neural network takes a pedestrian crossing the street for a whirlwind, there could be serious consequences.

Deep learning depends in an essential way on human expertise. It needs a huge dataset of cases prepared by humans in order to "beat" the humans at their own game (chess, go/weichi, etc). A world in which humans don't exist (or don't collaborate) would be a difficult place for deep learning. A world in which the expertise is generated by other deep-learning machines would be even tougher. For example, Google's translation software simply learns from all the translations that it can find. If many English-to-Italian human translators over the centuries have translated "table" as "tavolo", it learns to translate "table" into "tavolo". But what if someone injected into the Web thousands of erroneous translations of "table"? Scientists at Google are beginning to grapple with the fact that the dataset of correct translations, which is relentlessly being updated from what Google's "crawlers" find on the web, may degrade rapidly as humans start posting approximate translations made with... Google's translation software. If you publish a mistake made by the robot as if it were human knowledge, you fool all the other robots who are trying to learn from human expertise. Today's robots, equipped with deep learning, learn from our experts, not from each other. We learn from experts and by ourselves, i.e by "trial and error" or through a lengthy excruciating research. Robots learn from experts, human experts, the best human experts. Google's translation software is not the best expert in translation. If it starts learning from itself (from its own mediocre translations), it will never improve.

Supervised learning is "learning by imitation", which is as good as the person you are imitating. That's why the generation of AlphaGo is introducing additional tricks. Reinforcement learning, which was the topic of Minsky's PhD thesis in 1954, is a way to improve the speed and quality of machine learning. Another useful addition to deep learning (also used by AlphaGo) is tree-search, invented by Minsky's mentor Claude Shannon in 1950.

Similar considerations apply to robots. World knowledge is vital to perform ordinary actions. Robot dexterity has greatly improved thanks to a multitude of sensors, motors and processors. But grabbing an object is not only about directing the movement of the hand, but also about controlling it. Grabbing a paper cup is not the same as grabbing a book: the paper cup might collapse if your hand squeezes it too much. And grabbing a paper cup full of water is different from grabbing an empty paper cup: you don't want to spill the water. Moving about an environment requires knowledge about furniture, doors, windows, elevators, etc. The Stanford robot that in 2013 was trained to buy a cup of coffee at the cafeteria upstairs had to learn that a) you don't break the door when you pull down the handle; b) you don't spill coffee on yourself because it would cause a short circuit; c) you don't break the button that calls the elevator; etc; and, as mentioned, that the image in the elevator's mirror is you and you don't need to wait for yourself to come out of the elevator.

We interact with objects all the time, meaning that we know what we can do with any given object.

Your body has a history. The machine needs to know that history in order to navigate the labyrinth of your world and the even more confusing labyrinth of your intentions.

Finally, there are ethical principles. The definition of what constitutes "success" in the real world is not obvious. For example: getting to an appointment in time is "good", but not if this implies running over a few pedestrians; a self-driving car should avoid crashing against walls, unless it is the only way to avoid a child...

Most robots have been designed for and deployed in structured environments, such as factories, in which the goal to be achieved does not interfere with ordinary life. But a city street or a home contains much more than simply the tools to achieve a goal.

"Computers are useless: they can only give you answers" (Pablo Picasso, 1964).

We actually don't Think

The most successful algorithms used in the 2010s to perform machine translation use statistical analyses and require virtually no linguistic knowledge. These programs simply explore thousands of translations done by human experts and calculate which is the most popular. The very programmer who creates and improves the automatic-translation system doesn't need to have any knowledge of the two languages being translated into each other: it is only a statistical game. I doubt that this is how human interpreters translate one language into another, and i doubt that this approach will ever be able to match human translations, let alone surpass them.

Donald Knuth's famous sentence that A.I. seems better at emulating "thinking" than at emulating the things we do without thinking is still true; and it contains a larger truth. The really hard problem is that we don't know how we do the vast majority of things that we do, otherwise philosophers and psychologists would not have a job. A conversation is the typical example. We do it effortlessly. We shape strategies, we construct sentences, we understand the other party's strategy and sentences, we get passionate, we get angry, we try different strategies, we throw in jokes and we quote others. Anybody can do this without any training or education. And now, by comparison, check what kind of conversation can be carried out by the most powerful computer ever built.

Most of the things that we do by "thinking" (such as proving theorems and playing chess) can be emulated with a simple algorithm (especially if the environment around us has been shaped by society to be highly structured and to allow only for a very small set of moves). The things that we do without thinking cannot be emulated with a simple algorithm, if nothing else because even we don't know how we do them. We can't even explain how children learn in the first place.

Mind Uploading and Digital Immortality

Of all the life extension technologies proposed so far, perhaps none has captured the imagination of the machine-intelligence crowd more than mind uploading. Somehow the connection between the Singularity and digital immortality was made: at some point those super-intelligent machines will be able to perform one great task for us, upload our entire self and "become" us. Couple it with the immortality of the "cloud" (see later), and your "self" becomes immortal. It will be downloaded and uploaded from one release of the Singularity to the next one for the rest of time.

In the most memorable of Isaac Asimov's short stories, "The Last Question" (1956), humankind is preserved in cyberspace after the end of the universe. Some forms of mind uploading already appeared in Arthur Clarke's novella "The City and the Stars" (1953) and in Frederick Pohl's short story "The Tunnel Under the World" (1955). I find the latter more realistic because it envisions mind uploading as a trick devised by the advertising industry.

The technology of uploading a human mind into a computer was first explored by a geneticist, George Martin, in "A Brief Proposal on Immortality" (1971). He foresaw that someday computers would become so powerful that they will be able to do everything that a brain can do. Therefore why not simply port our brains to computers and let the computers do the job. Needless to say, philosophers are still arguing whether that "mind" would still be "me" once uploaded into software instead of gray matter. Hans Moravec speculated that you are just a pattern, therefore you could "transmigrate" to a different body ("Dualism Through Reductionism", 1986).

That vision became more realistic in the 1990s with the explosion of the World-wide Web. A paleontologist, Gregory Paul, in collaboration with a mathematician, Earl Cox, speculated about cyber-evolution that could create non-human minds in "Beyond Humanity" (1996), including the idea of immortal "brain carriers" to replace our mortal bodies. In the days when television was still influential, William Gibson, the science-fiction writer who a decade earlier had invented the term "cyberspace" ("Burning Chrome", 1982), contributed to the popularization of the concept by scripting an X-Files episode, "Kill Switch" (1998), in which a man uploads his mind into cyberspace. Ray Kurzweil wrote the article "Live Forever Uploading The Human Brain" (2000).

Then came the deluge with books such as Richard Doyle's "Wetwares - Experiments in PostVital Living" (2003) exploring all sorts of technologies of immortality. Every year the vision of what Martine Rothblatt calls "mindclones", implemented in "mindware" (the software for consciousness), has to be updated to the latest computer platform.

In 2012 a Russian tycoon, Dmitry Itskov, pretty much summarized the vision of the immortality field: firstly, build brain-machine interfaces so that a human brain can control a robotic body; secondly, surgically transplant the human brain into a robotic body; and, finally, find a method to achieve the same result without the gory surgical operation, i.e. a way to upload a person's mind into the robotic body or, for that matter, into just about anything.

The question, of course, is whether that "you" will that still be you. Or just a machine mimicking you? You can ask someone to impersonate you, but that does not mean that he or she is you. That someone has absorbed the "pattern" of your behavior, but s/he is not you. By the same token, if a machine absorbs some pattern found in your brain, it doesn't mean that the machine has become you. We literally don't know what in the brain makes you "you" (and not, for example, me). This disembodied and reconstituted "mind" might well be immortal, but is it you?

In other words, this program is predicated on the assumption that "i" am entirely in my brain, and that my body is simply a vehicle for my "i" to survive. If so, such body can as well be replaced by some other material substrate. The brain is disposable, according to this view: the brain is merely the organ of the body designated to host the processes that construct the "i", but the "i" truly is only those processes, which, luckily for us, turn out to be information-based processes, which, luckily for us, can be easily transplanted from the mortal (and, let's admit it, quite repulsive) brain into the kind of information-processing machines that we started building in the 1940s and that are getting more and more powerful, rapidly approaching the capacity required to simulate the entirety of those brain processes.

This movement has revived the project of whole brain emulation. Ray Kurzweil and others have estimated that "artificial general intelligence" will be achieved first via whole brain emulation. The basic idea is to construct a complete detailed software model of a human brain so that the hardware connected to that software will behave exactly like the human would (which includes answering the question "is it really you?" with a "yes").

But first one needs to map the brain, which is not trivial. In 1986 John White's and Sydney Brenner's team mapped the brain of the millimeter-long worm Caenorhabditis Elegans (302 neurons and 7000 synapses). As far as i know, that is still the only brain that we have fully mapped. And it took twelve years to complete that relatively simple "connectome". The term was coined in Olaf Sporns' "The Human Connectome, a Structural Description of the Human Brain" (2005). A connectome is the map of all the neural connections in a brain. In 2009, a few years after the success of the Human Genome Project, the USA launched the Human Connectome Project to map the human brain. The task, however, is not on the same scale as mapping a worm's brain. The entire human genome is represented by about a few gigabytes of data. Cellular biologist Jeff Lichtman and Narayanan Kasthuri estimated that a full human connectome would require one trillion gigabytes of memory ("Neurocartography", 2010). Furthermore, we all share (roughly) the same genome, whereas each brain is different. The slightest mistake and... oops... they may upload the brain of someone else instead of yours.

Once we are able to map brains, we will need to interface those brains with machines. This may actually come sooner. In 1969 the Spanish neurophysiologist Jose Delgado implanted devices in the brain of a monkey and then sent signals in response to the brain's activity, thus creating the first bidirectional brain-machine-brain interface. In 2002 John Chapin debuted his "roborats", rats whose brains were fed electrical signals via a remote computer to guide their movements. His pupil Miguel Nicolelis achieved the feat of making a monkey's brain control a robot's arm. In 2008 the team made the monkey control a remote robot (in fact, located in another continent).

By the time science is capable of uploading your mind to cyberspace most of us will probably be dead, and with us our brains. That disturbing thought predated the very science we are talking about. Robert Ettinger's book "The Prospect of Immortality" (1962) is considered the manifesto of "cryonics", the discipline of preserving brains by freezing them. It was actually cryonics that started the "life extension" movement. In 1964 another founding father, Evan Cooper, launched the Life Extension Society (LES). In 1972 Fred Chamberlain, a space scientist at the Jet Propulsion Laboratory, founded the Alcor Society for Solid State Hypothermia (ALCOR), now called Alcor Life Extension Foundation, to enter that business.

The similarities with the most successful organized religions of the Western world are too obvious to be overlooked. The end of the world is coming in the form of the Singularity, but, not to worry, we will all be resurrected in the form of mind uploads made possible by the super-machines of that very Singularity. The only difference with the ancient Western religions is that people from previous ages are dead for good, forever: we don't have their brains to upload anymore. But then maybe those super-human machines will find a way to resurrect the dead too.

Machine Immortality and the Cloud

The other implicit assumption in the scenario of mind uploading is that these superhuman machines, capable of self-repairing and of self-replicating, will live forever.

That would represent a welcome change from what we are used to. The longest life expectancy for an electrical machine probably belongs to refrigerators, that can last longer than a human generation. Most appliances die within a decade. Computers are the most fragile of all machines: their life expectancy is just a few years. Their "memories" last less than human memory: if you stored your data on a floppy disc twenty years ago, there is probably no way for you to retrieve them today. Your CDs and DVDs will die before you. And even if your files survived longer, good luck finding an application that can still read them. Laptops, notepads and smartphones age increasingly faster. The life expectancy of machines seems to be decreasing ever more rapidly. And, of course, they are alive only for as long as they are plugged into an electrical outlet (battery life can be as little as a few hours); and they seem to be more vulnerable than humans to "viruses" and "bugs".

One has to be an inveterate optimist to infer from the state of the art in storage media that increasingly mortal and highly vulnerable computer technology is soon to become immortal.

Of course, the Singularity crowd will point out the "cloud", where someone else will take care of transferring your data from one dying storage to a newer one and translating them from one extinct format to a newer one. Hopefully some day the cloud will achieve, if not immortality, at least the reliability and long lifespan of our public libraries, where books have lasted millennia.

Having little faith in software engineers, i am a bit terrified at the idea that some day the "cloud" will contain all the knowledge of the human race: one little "bug" and human civilization as we know it will be wiped out in a second. It is already impressive how many people lose pictures, phone numbers and e-mail lists because of this or that failure of a device. All it takes is that you forget to click on some esoteric command called "safely remove" or "eject" and an entire external disc may become corrupted.

If and when super-intelligent machines come, i fear that they will come with their own deadly viruses, just like human intelligence came (alas) with the likes of influenza pandemics, AIDS (Acquired ImmunoDeficiency Syndrome), SARS (Severe Acute Respiratory Syndrome), Ebola and Zika. And that's not to mention the likelihood of intentional cyber terrorism (i'm not sure who's getting better at it: the cryptographers who are protecting our data or the hackers who are stealing them) and of "malware" in general. If today they can affect millions of computers in a few seconds, imagine what the risk would be the day that all the knowledge of the world is held in the same place, reachable in nanoseconds. The old computer viruses were created for fun by amateurs. We are entering the age in which "cyber crime" will be the domain of super-specialists hired by terrorists and governments. Originally, a computer virus was designed to be visible: that was the reward for its creator. Today's cyber crime is designed to be invisible... until it's too late.

Remember when only fire could destroy your handwritten notes on paper? And it was so easy to make photocopies (or even manual copies) of those handwritten notes? We found the Dead Sea Scrolls two thousand years after they had been written (on a combination of vellum and papyrus), and the Rosetta stone is still readable after 2,200 years. I wonder how many data that we are writing today will still be found two thousand years from now in the "cloud".

Hackers will keep getting more and more sophisticated and, when armed with powerful computers provided by rich governments, able to enter any computer that is online and access its contents (and possibly destroy them). In the old days the only way for a spy to steal a document was to infiltrate a building, search it, find the safe where the documents were being held, crack open the safe or bribe someone, duplicate the documents, flee. This was dangerous and time consuming. It could take years. Today a hacker can steal thousands if not millions of documents while comfortably sitting at her desk, and in a fraction of a second. The very nature of digital files makes it easy to search and find what you are looking for.

Ironically, an easy way to make your files safe from hacking is to print them and then delete them from all computers. The hacker who wants to steal those files is now powerless, and has to be replaced by a traditional thief who has to break into your house, a much more complicated proposition.

Cyber-experts now admit that anything you write in digital form and store on a device that is directly or indirectly connected to the Internet, will, sooner or later, be stolen. Or destroyed. When, in march 2013, the websites of JPMorgan Chase and then American Express were taken offline for a few hours after being attacked by the Izz ad-Din al-Qassam Cyber Fighters, cyber-security expert Alan Paller warned that cyber-attacks are changing from espionage to destruction. A malware to destroy (digital) information on a large scale would be even easier to manufacture than the malware Stuxnet (unleashed in 2010 probably by Israel and the USA) that damaged about one thousand centrifuges used to enrich nuclear material at Iran's nuclear facility in Natanz.

I also feel that "knowledge" cannot be completely abstracted from the medium, although i find it hard to explain what the difference is between knowledge stored in Socrates' mind, knowledge stored in a library and knowledge stored in a "cloud". A co-founder of one of the main cloud-computing providers said to me: "Some day we'll burn all the books". Heinrich Heine's play "Almansor", written a century before Adolf Hitler's gas chambers, has a famous line: "Where they burn books, they will ultimately burn people too". Unlike most predictions about machine intelligence, that is one prediction that came true.

Corollary: Digital Media Immortality

If you want to turn yourself into data, instead of flesh and bones, and hope that this will make you immortal, you have a small technical problem to solve.

As you well know from your Christmas shopping, the capacity of computer storage media (for the same price) increases rapidly. That's the good news. The bad news is that its longevity has been decreasing, and significantly decreasing if you start from way back in time. The life expectancy of paper and ink is very long in appropriate conditions. The original storage media for computers, punched paper tapes and punch cards, are still readable 70 years later: unfortunately, the machines that can read them don't exist anymore, unless you have access to a computer museum. By comparison, the life expectancy of magnetic media is very, very short. Most people born before 1980 have never seen a magnetic tape except in old sci-fi movies. It was introduced with the first commercial computer, the Eckert-Mauchly's UNIVAC I, in 1951. Today most magnetic tapes store terabytes of data. They last about 20-30 years. Nobody knows how long the multiplatter disks from the mainframes of the 1960s lasted because they got out of fashion before we could test their lifespan. Floppy discs are magnetic disks, the most common type of which had a capacity of 1.44 megabytes or 2 megabytes. The 8" floppy disks of the 1970s and the 5.25" floppy disks of the 1980s are given a life expectancy of 3-5 years by those who never used them, but those like me who still have them know that at least half of them are still working 30 years later. The external "hard disks" that replaced them (and that today can easily hold a terabyte, i.e. a million times more data) may last longer, but they need to spin in order to be read or written, and spinning-disk hard drives don't last long: they are mechanical devices that are likely to break long before the magnetic layer itself deteriorates, especially if you carry them around (in other words, if you use them).

Music was stored on magnetic tapes, and later on cassettes, that would still work today if mass-market magnetic tape players still existed, although they would probably not sound too good, and on vinyl records, that definitely still play today if you didn't scratch them and used appropriate cartridges on your turntable like i did. My cassettes from the 1970s still play ok. Video was stored on VHS tapes, that still play today (i have about 300 of them), but, again, colors and audio may not look/sound so good after years of playing on a VCR (if you can still find a VCR).

Then came the optical generation. Rewritable optical discs are much less reliable for data storage than read-only optical discs that you buy/rent at music or video stores because they are physically made of different materials (the film layer degrades at a faster rate than the dye used in read-only discs). The jury is still out on optical media, but, as far as storing your data goes, the Optical Storage Technology Association (OSTA) estimates a lifespan of 10-25 years for compact discs (CDs), that typically held 650 megabytes (or the equivalent of 700 floppy disks), and digital video discs (DVDs), that typically held 4.7 gigabytes. However, in practice, optical devices are much more likely to get damaged because very few people store their discs in optimal conditions. Just leaving them on a desk unprotected may greatly shorten their lifespans just like anything else that you look at (optical is optical).

Now we live in the age of solid-state media, devices that don't have moving parts and that can store several gigabytes on a very small device, like USB flash drives ("thumb" drives) and secure-digital cards ("flash cards"). They are generally less (not more) reliable than hard drives, and the manufacturers themselves don't expect them to last longer than about eight years.

And that's not to mention the quality of the recording: digital media are digital, not analog. You may not be able to tell the difference because your ears are not as good as the ears of many (supposedly less intelligent) animals, but the digital music on your smartphone is not as accurate a recording as the vinyl record of your parents or the 78 RPM record of your grandparents. A digital recording loses information. The advantage, in theory, is that the medium is less likely to deteriorate as you use it: magnetic tape degrades every time it passes by the magnetic head of a cassette player or a VCR, and the grooves of LPs do not improve when the cartridge of the turntable rides on them. The advantage of the old media, however, is that they "degraded": they didn't simply stop working. Digital files are either perfect or don't work, period. My old VHS tapes lost some of the color and audio fidelity, but i can still watch the movie. Many of my newer DVDs stop in the middle of the movie, and there is no way to continue. (I am also greatly annoyed by the difficulty of rewinding/forwarding a DVD or pinpointing a frame of the movie, something that can easily be done on a VHS tape: this is possibly the first "regress" in history for random access, a feature introduced by the Romans when they switched from the scroll to the codex).

On the other hand, microfilms are estimated to last 500 years: that is a technology that was introduced by John Benjamin Dancer in 1839, and first used on a large scale in 1927 by the Library of Congress of the USA (that microfilmed millions of pages in that year).

You can tell that the plot remains the same: larger and larger storage, but perhaps less and less reliable.

Note that all of this is very approximate: search for the longevity of free neutrons, and you'll readily find it (14'42"), but if you search for a scientific answer to the question of storage media longevity, you will not find it. That's how advanced the science of storage media is.

Finally, even if your media could potentially last a long time, when is the last time you saw a new computer model with a floppy drive? Even optical drives (CD, DVD) are disappearing as i type these words, and your favorite flash memory may become obsolete before this book goes out of print. And even if you still find a machine with a drive for your old media, good luck finding the operating system that has a file system capable of reading them. And even if you find both the hard drive and the operating system that can read them, good luck finding a copy of the software application that can read the data on them (e.g., GEM was the popular slide presentation software in the heydays of floppy discs). This is a field in which "accelerating progress" (in physical media, operating systems and viewing applications) has been consistently hurting data longevity, not extending it.

Yes, i know: the "cloud" will solve all these problems. And create bigger ones. What will happen if some day the electrical grid shuts down for a few weeks?

Ethical Intermezzo: The Moral Consequences of Human Immortality

If immortality can be achieved in this life, it will have non-trivial consequences on a selfish race like the human race.

If you believe that immortality is granted in the afterlife, you will do everything that you can in order to obtain it in the afterlife (which typically means obeying the instructions that a god gave humans to achieve the above said immortality); but if you believe that immortality is granted in this life, you will do everything that you can to obtain it in this life. A person who believes that immortality will be granted in the afterlife based on her good deeds will promptly sacrifice her life to save someone else or to fight a dangerous disease in Africa or to provide her children with a better future; but a person who believes that immortality is granted in this life has literally no motivation to risk her life to save someone else's life, nor any motivation to risk her life in Africa nor (ultimately) any motivation to care for her own children. Once you are dead, you are dead, therefore the only meaning that your life can have is not to die. Never. Under no circumstances. No matter what. Stay alive.

The new morality of a society that believes in immortality here now will be simple: stay alive at all costs, because immortality in this life is the only thing that matters.

"Humanity is acquiring all the right technology for all the wrong reasons" (Buckminster Fuller, "Earth Inc", 1973)

Another Philosophical Intermezzo: Do We Really Want Intelligence at All?

Intelligence is messy. When we interact with human beings, we have to consider their state of mind besides our immediate goal. We may only need a simple favor, but it makes a huge difference whether our interlocutor is happy or sad, on vacation or asleep, has just lost a close relative or been injured in an accident, angry at us, busy with her work, etc. Whether our interlocutor is capable or not of performing that favor for us may be a secondary factor compared with whether she is in the mental condition of doing it and doing it right now. On the other hand, when we deal with a dumb machine, the only issue is whether the machine is capable of performing the task or not. If it is, and the power chord is plugged into the power outlet, it will. It won't complain that it's tired or in a bad mood, it won't ask us for a cigarette, it won't spend ten minutes gossiping about the neighbors, it won't comment on the government or the soccer game.

It may seem a paradox, but, as long as machines are dumb, they are easy and painless to interact with. They simply do what we ask them to do. No whims. No complaints. No formalities.

The complication that comes with intelligent beings is that they are subject to moods, feelings, opinions, intentions, motives, etc. There is a complicated cognitive apparatus at work that determines the unpredictable reaction of an intelligent being when you ask even the simplest of questions. If your wife is mad at you, even the simplest question "What time is it?" might not get an answer. On the other hand, if you use the right manners at the right time, a complete stranger may do something truly important for you. In many cases it is crucial to know how to motivate people. But in other cases that is not enough (if the person is in a bad mood for reasons that are totally independent of your will). Human beings are a mess. Dealing with them is a major project. And that's not to mention the fact that human beings sleep, get sick, go on vacation, and even take lunch breaks. In Western Europe they are often on strike.

Compare humans with dumb machines that simply do what you ask. For example, the automatic teller machine hands you money at any time of the day or night any day of the year. Wherever the intelligent being has been replaced by a dumb machine, the interaction is simpler. We structured the interaction so that the dumb machine can perform all the operations that we need.

The reason that automated customer support has replaced human beings in so many fields is not only that it is cheaper to operate by the provider but also that it is preferred in the majority of cases by the majority of customers. The honest truth is that very few of us enjoy waiting for an operator to greet us "Hello? How are you? Isn't it a beautiful day? How can i help you?" Most of us prefer to press digits on a phone keypad. The truth is that most customers are happy if we remove the complication of dealing with human beings.

When i worked in the corporate world, my top two frustrations were the secretary and the middle management. Dealing with the secretary (especially in unionized Italy) required superior psychological skills: say the wrong word in the wrong tone, and s/he'd boycott you for the rest of day. Most middle managers were mediocre and mostly slowed down things, seemingly paid mainly to kill great ideas. The only way to get important things done quickly was, again, to use the art of psychology: befriend them, chat with them, find out what motivated them, offer them rides home, hang out with them. My life would have been much easier if my colleagues and my secretary had been heartless robots.

And, let's face it, we often don't have the patience for human interactions that involve protocols of behavior. We are often happy when good manners are replaced by cold mechanic interactions shaped by goals and restrained by laws. Hence we do not really want machines with human intelligence, i.e. we don't want them to have emotions, to be verbose, to deceive, to plead, etc. One goal of inventing machines is precisely to remove all of that, to remove that inefficient, annoying, time-consuming quality of "humanity".

We removed the human/intelligent element from many facets of ordinary life because the truth is that in most cases we don't want to deal with intelligent beings. We want to deal with very dumb machines that will perform a very simple action when we press a button.

I'll let psychologists and anthropologists study the reasons for this trend towards less and less human interactions, but the point here is that intelligence comes at a high price: intelligence comes with feelings, opinions, habits, and a lot of other baggage. You can't have real intelligence without that baggage.

When we study how to create intelligent machines, do we really mean "intelligent" or do we mean "stupid in a way that will serve our intelligence"?

Religion and the Law of Accelerated Exaggeration

Robert Geraci of Manhattan College, in his book "Apocalyptic A.I." (2010), showed that Singularity thinking borrows motifs and practices from Jewish and Christian apocalyptic scriptures. The Judeo/Christian religions offer a dualistic view of the world: good and evil are fighting a cosmic battle. "Evil" is manifested as bodily decay, earthly world, and limited intellect. "Good" will someday materialize as eternal life, celestial world and unlimited knowledge. Singularity thinking (which he calls "Apocalyptic A.I.") adopts a similar view, cursing the mortal body and the limited knowledge of the human mind while envisioning a future in which we will become immortal and omniscient in cyberspace. The enabler is the high priesthood of A.I. scientists and engineers, whom Geraci nicknames "mystical engineers".

Geraci points out how apocalyptic thinking arose among Jews and Christians: they were both persecuted people. The Jews endured slavery and/or occupation by the Assyrians, the Babylonians, the Greeks and the Romans. The Christians were persecuted by the Romans. Geraci thinks that A.I. scientists such as Hans Moravec and Ray Kurzweil feel similarly persecuted, except that now it is "bodily alienation": they want to escape the limitations of the biological body.

New York University anthropologist Stefan Helmreich in "Silicon Second Nature" (1998) studied the "mystical" attitudes of the practitioners of Virtual Reality and Artificial Life. In 2003 Philip Rosedale's Linden Lab launched "Second Life", a virtual world accessible via the Internet in which a user could adopt a new identity and live a "second life" as an avatar, and Geraci views Second Life as a sort of temple where people perform religious functions.

Incidentally, the Singularity bears obvious similarities with the Omega Point, described by Pierre Teilhard, a Catholic priest from France, in his book "The Phenomenon of Man" (1955), and conceived as a point of super-human intelligence towards which the universe is evolving. The physicist Frank Tipler gave the omega point a formal mathematical and scientific formulation in his book "The Physics of Immortality" (1994).

Cultural historian Margaret Wertheim in "The Pearly Gates of Cyberspace" (1999) argued that cyberspace represents the high-tech equivalent of religious paradise, an identification that goes back to Michael Benedikt of the University of Texas at Austin, who wrote in the introduction to the anthology "Cyberspace" (1992) that cyberspace is the equivalent of the biblical "Heavenly City". Jeffrey Fisher's study "The Postmodern Paradiso" (1997) shows similarities between medieval Christian mythology and cyberspace utopia. To celebrate the age of Web 2.0, in 1999 Yale University's computer scientist David Gelernter wrote a manifesto titled "The Second Coming".

The impact on mysticism of the discovery of cyberspace has not been too different from the impact that the discovery of America had five centuries earlier: when in 1503 Amerigo Vespucci wrote to Lorenzo de Medici that Cristoforo Colombo had actually discovered a "new world", many viewed this "New World" as the new Eden. America (named after Amerigo) became the natural vehicle for Europe's utopian dreams at a time when Europe was launching into the humanistic, scientific and artistic revolutions of the Renaissance. The great utopian works of the following century, from Thomas More's Utopia" (1516) to Francis Bacon's "New Atlantis" (1627), were influenced by the myth of America as a blank space where a superior society could be created. Six centuries later what today's futurists are imagining in cyberpace is not all too different from what those 16th-17th century futurists imagined in their utopian books.

In "The Future of Religion" (1985) sociologists Rodney Stark of the University of Washington and William Bainbridge of Boston University argued that secularism encourages religious innovation rather than signalling the outright demise of religion. In other words, an increasingly secular science has not killed religion but rather has created an opportunity for reforming religion. When Nietzsche announced "the death of God" in his book "Thus Spoke Zarathustra" (1883), he had basically opened the doors for a new religion, and the first one to take advantage of that opening had been the scientistic religion presented by Karl Marx in "Capital" (1894): communism. In his "Religions for a Galactic Civilization" (1982) Bainbridge advocated establishing a scientistic theocracy along the lines of UFOlogy as something that humans need in order to survive (UFOlogy was replaced by Singularity thinking in the revised 2009 version).

Wertheim thinks that humans naturally want a spiritual dimension to their lives. Science, by banning the spiritual out of the physical universe, has created the need for a new kind of spiritual space. If they can no longer find it in the physical universe, today's humans will find it in cyberspace. Virtual life on the Internet has been getting more and more interesting and meaningful, and the line between the real world and the virtual world has gotten more and more blurred.

Bainbridge wrote in "Religion for a Galactic Civilization 2.0" (2009) that religion and science are not opposed at all; instead, they coevolve: "Religion shapes science and technology, and is shaped by them in return". And, without mentioning the Singularity, he added that the "creation of a galactic civilization may depend upon the emergence of a galactic religion capable of motivating society for the centuries required to accomplish that great project".

Traditionally the strength of religion has been inversely proportional to the status of science in society. But this time the new religion of A.I. is about science itself and it is being created by people who are very knowledgeable about the science. This is not the first time that scientists present technology as a sort of divine power, as David Noble of York University in Toronto has shown in "The Religion of Technology" (1997), and it would not be the first time that a new science rises in parallel with a new organized religion, as Wertheim has shown in "In Pythagoras' Trousers" (1997). Francis Bacon's "New Atlantis" (1627) was the first scientific utopia, and Isaac Newton wrote (unpublished) books of prophecy such as "Observations upon the Prophecies of Daniel, and the Apocalypse of St John" (1733). Sometimes we forget that science and technology evolved from the Catholic monasteries and from the Church-controlled universities (and from the Islamic madrasas) of the Middle Ages. The culture of the San Francisco Bay Area lies at the same intersection of science and spirituality, the former represented by Silicon Valley's high-tech industry and the latter by the "New Age" movement. Fred Turner calls it "digital utopianism" in his book "From Counterculture to Cyberculture" (2008).

Just like prophetic books mediated between science and religion back then, today it is science fiction that has mediated between religion and technology. Critical studies such as David Ketterer's "New Worlds for Old" (1974) showed that science fiction routinely borrows concepts from the Christian scriptures. Studies such as Thomas Disch's "The Dreams Our Stuff Is Made Of" (1998) and Jason Pontin's "On Science Fiction" (2007) documented how science fiction exerted a huge influence on A.I. scientists. Pontin once wrote "Science fiction is to technology as romance novels are to marriage: a form of propaganda" (MIT Technology Review, 2005). Many future A.I. scientists were inspired to enter the A.I. field precisely because they were fans of science fiction: Isaac Asimov's "I Robot" stories of the 1940s and "Multivac" stories of the 1950s, Osamu Tezuka's manga "Tetsuwan Atomu/ Astro Boy" (1951), Arthur Clarke's short story "Dial F for Frankenstein" (1964), Frank Herbert's "Do I Wake or Dream/ Destination Void" (1965), Brian Aldiss' short story "Super-Toys Last All Summer Long" (1968), Philip Dick's "Do Androids Dream of Electric Sheep" (1968), Algis Budrys' "Michaelmas" (1977), Douglas Adams' "The Hitchhiker's Guide to the Galaxy" (1979), Vernon Vinge's novella "True Names" (1981), William Gibson's "Neuromancer" (1984), which was predated by his short story "Burning Chrome" (1982), Neal Stephenson's "Snow Crash" (1991), etc. After all, even the great theoretical physicist Freeman Dyson wrote in his visionary book "Imagined Worlds" (1998) that "science is my territory, but science fiction is the landscape of my dreams".

Science fiction inspired the "transhumanist" movement way before the Singularity became a popular concept. The Extropian movement believed in the power of science and technology to yield immortality. Its members practiced cryogenics to preserve their brain after death. The term "extropy" was coined by Tom Bell, juxtaposing it to "entropy". The Oxford philosopher Max More had helped set up the first cryonic service in Europe (later renamed Alcor). Relocating to Los Angeles, in 1988 More started the magazine Extropy, subtitled "journal of transhumanist thought" and founded the Extropy Institute, which in 1991 had its own online forum. The Extropian movement had strong anti-government libertarian/anarchic political views, predicting a technocratic society in which power would be wielded directly by the people. By the time Wired published the influential article "Meet The Extropians" in 1994, the extropian movement included members and sympathyzers such as Hans Moravec, Ralph Merkle, Nick Szabo, Hal Finney, as well as co-founders Tom Bell (Tom Morrow) and Perry Metzger. Merkle would go on to become a leader in nanotechnology, Szabo and Finley would pioneer Bitcoin, Metzger would launch the cryptography mailing-list.

Nick Bostrom, a Swedish philosopher at Oxford University, has pursued more social and ethical concerns in the several organizations that he established: in 1998 Bostrom and fellow philosopher David Pearce founded the World Transhumanist Association that later changed name to Humanity+, and in the same year Bostrom published "How Long Before Superintelligence?" (1998). In 2004 Bostrom and James Hughes founded the Institute for Ethics and Emerging Technologies; and in 2005 Bostrom founded the Future of Humanity Institute at Oxford University.

Kevin Kelly explored the connection between information and God in the article "Nerd Theology" (1999).

In 2006 the Italian physicist Giulio Prisco became an advocate in virtual reality for the transhumanist movement, initially through his avatar Giulio Perhaps in Second Life. In 2007 he published the article "Engineering Transcendence" predicting that in the future it will be possible to become immortal inside cyberspace and to create perfect simulations of the past that will revive all those who have ever been alive. In 2008 he founded the Order of Cosmic Engineers (in a virtual world) and in 2010 the Turing Church (in the real world). The latter was initially just a "mailing list about the intersection of transhumanism and spirituality", but in 2014 it evolved into a "minimalist, open, extensible" religion whose manifesto preaches: "We will go to the stars and find Gods, build Gods, become Gods, and resurrect the dead from the past with advanced science"

These "un-religions" (religions with neither a hierarchy of priests nor immutable dogmas) are reminiscent of the church of engineers envisioned by August Comte, the founder of positivism, in his book "Catechism of Positive Religion" (1852). Comte was hoping to replace all religious institutions (in his view outdated) with a scientistic religion.

"There are things known and there are things unknown, and in between are the doors of perception" (Aldous Huxley).

Analog vs Digital

Most machines in the history of human civilization were and are analog machines, from the waterwheel to your car's engine. A lot of the marvel about computers comes from the fact that they are digital devices: once you digitize texts, sounds, images, films and so forth the digital machine can perform, at incredible speed, operations that used to take armies of human workers or specialists armed with expensive specialized machines. Basically, digitizing something means reducing it to numbers, and, therefore, the mind-boggling speed at which computers perform calculations, gets automatically transferred to other fields, such as managing texts, audio and video. The "editing" feature is, in fact, one of the great revolutions that came with the digital world. Previously, it was difficult and time-consuming to edit anything (text, audio, photos, video). Filing, editing and transmitting are operations that have been dramatically revolutionized by progress in digital technology and by the parallel process of digitizing everything, one process fueling the other.

Now that television broadcasts, rented movies, songs and books are produced and distributed in digital formats, i wonder if people of the future will even know what "analog" means. Analog is any physical property whose measurable values vary in a continuous range. Everything in nature is analog: the weight of boulders, the distance between cities, the color of cherries, etc. (At microscopic levels nature is not so analog, hence Quantum Theory, but that's another story). Digital is a physical property whose measurable values are only a few. The digital devices of today can typically handle only two values: zero and one. Actually, i don't know any digital device that is not binary. Hence, de facto, in our age "digital" and "binary" mean the same thing. Numbers other than zero and one can be represented by sequences of zeroes and ones (e.g. a computer internally turns 5 into 101). Texts, sounds and images are represented according to specific codes (such as ASCII, MP3 and MP4) that turn texts, sounds and images into strings of zeroes and ones.

The easiest way to visualize the difference between analog and digital is to think of the century-old bell-tower clocks (with the two hands crawling between the 12 Roman numerals) and the digital clock (that simply displays the time in hours/minutes).

When we turn a property from analog to digital we enable computers to deal with it. Therefore you can now edit, copy and email a song (with simple commands) because it has been reduced to a music file (to a string of zeroes and ones).

Audiophiles still argue whether digital "sounds" the same as analog. I personally think that it does (at today's bit rates) but the stubborn audiophile has a point: whenever we digitize an item, something is lost. The digital clock that displays "12:45" does not possess the information of how many seconds are missing to 12:46. Yesterday's analog clock contained that information in the exact position of the minute hand. That piece of information may have been useless (and obtainable only by someone equipped with a magnifying glass and a pocket calculator) but nonetheless the device had it. The music file is not an exact replica of the song: when the musicians performed it, they were producing an analog object. Once that analog object is turned into a digital file, an infinite number of details have been lost. The human ear is limited and therefore won't notice (except the above said stubborn audiophiles). We don't mind because our senses can only experience a limited range of audio and visual frequencies. And we don't mind also because amazing features become available with digital files, for example, the ability to improve the colors of a photograph so we can pretend that it was a beautiful vacation when in fact it rained all the time.

When machines carry out human activities, they are "digitizing" those activities; and they are digitizing the "mental" processes that lie behind those activities. In fact, machines can manage those human activities only after humans digitized (turned into computer files) everything that those human activities require, for example maps of the territory.

Using digital electronic computers to mimic the brain is particularly tempting because it was discovered that neurons work like on/off switches. They "fire" when the cumulative signal that they receive from other neurons exceeds a certain threshold value, otherwise they don't. Binary logic, invented in 1854 by the British philosopher George Boole in a book titled "The Laws of Thought", seems to lie at the very foundation of human thinking. In fact, as early as 1943, Warren McCulloch, in cooperation with Walter Pitts, described mathematically an "artificial" neuron that can only be in one of two possible states. A population of artificial binary neurons can then be connected in a very intricate network to mimic the way the brain works. When signals are sent into the network, they spread to its neurons according to the simple rule that any neuron receiving enough positive signals from other neurons sends a signal to other neurons. It gets better: McCulloch and Pitts proved that such a network of binary neurons is fully equivalent to a Universal Turing Machine.

There is, however, a catch: McCulloch's binary neurons integrate their input signals at discrete intervals of time, rather than continuously as our brain's neurons do. Every computer has a central clock that sets the pace for its logic, whereas the brain relies on asynchronous signaling because there is no synchronizing central clock. If you get into the finer details of how the brain works, there are more "analog" processes at work, and there are analog processes inside the neuron itself (which is not just an on/off switch).

One could argue that the brain is regulated by the body's internal clocks (that regulate every function, from your heart to your vision) and therefore the brain behaves like a digital machine; and that everything is made of discrete objects all the way down to quarks and leptons; hence nothing in nature is truly analog. Even if you want to be picky and invoke Quantum Theory, the fact remains that a brain uses a lot more than zeroes and ones; a computer can only deal with zeroes and ones. As tempting as it is to see the brain as a machine based on binary logic, the difference between the human brain and any computer system (no matter how complex the latter becomes) is that a computer is way more "digital" than a brain. We know so little about the brain that it is difficult to estimate how many of its processes involve a lot more than on/off switching, but a safe guess is that there are several hundreds. Despite the illusion created by the McCulloch-Pitts neuron, a computer is a binary machine which the brain is not.

There might be a reason if a brain operates at 10-100 Hz whereas today's common microprocessors need to operate at 2-3 Gigahertz (billions of Hz), hundreds of millions of times faster, to do a lot less; also human brains consume about 20 watts and can do a lot more things than a supercomputer that consumes millions of watts. Biological brains need to be low-power consumption machines or they would not survive. There are obviously principles at work in a brain that have eluded computer scientists.

Carver Mead's "neuromorphic" approach to machine intelligence is not feasible for the simple reason that we don't know how the brain works. Based upon the Human Genome Project (that successfully decoded the human genome in 2003), the USA launched the "Brain Initiative" in April 2013 to map every neuron and every synapse in the brain.

There are also government-funded projects to build an electronic model of the brain: Europe's Human Brain Project and the USA's Systems of Neuromorphic Adaptive Plastic Scalable Electronics (SYNAPSE), sponsored by the same agency, DARPA, that originally sponsored the Arpanet/Internet. Both Karlheinz Meier in

Germany and Giacomo Indiveri in Switzerland are toying with analog machines. The signaling from one node to the others better mimics the "action potentials" that trigger the work of neurons in the human brain and requires much less power than the ones employed in digital computers. SYNAPSE (2008) spawned two projects in California, one run by Narayan Srinivasa at Hughes Research Laboratories (HRL) and the other run by Dharmendra Modha at IBM's Almaden Labs in Silicon Valley. The latter announced in 2012 that a supercomputer was able to simulate 100 trillion synapses from a monkey brain, and in 2013 unveiled its "neuromorphic" chip TrueNorth (not built according to the traditional John von Neumann architecture) that can simulate 1 million neurons and 256 million synapses. This represented the first building block to push computer science beyond the Von Neumann architecture that has ruled since the early days of electronic computation. Interestingly, this chip (consuming only 70 milliwatts of power) was also one of the most power-efficient chips in the history of computing... just like the human brain.

Analog Computation

Neural networks, and deep learning in particular, are good for recognizing patterns (e.g., that this particular object is an apple) but not for learning events in time. Neural networks have no sense of time.

In 1992 Hava Siegelmann of Bar-Ilan University in Israel and Eduardo Sontag of Rutgers University developed Recurrent Neural Networks (RNNs) that can operate on sequences and therefore can also model relationships in time (see: "Analog Computation via Neural Networks", a paper submitted in 1992 but published only in 1994). Typical applications of RNNs are: image captioning, that turns an image into a sequence of words ("sequence output"); sentence classification, that turns a sequence of words into a category ("sequence input"); and sentence translation (sequence input and sequence output). The innovation in RNNs is a hidden layer that connects two points in time. In the traditional feed-forward structure, each layer of a neural network feeds into the next layer. In RNNs there is a hidden layer that feeds not only into the next layer but also into itself at the next time step. This recursion or cycle adds a model of time to traditional backpropagation, and is therefore known as "backpropagation through time".

A general problem of neural networks with many layers ("deep" neural networks), and of RNNs in particular, is the "vanishing gradient", already described in 1991 by Josef "Sepp" Hochreiter at the Technical University of Munich and more famously in 1994 by Yoshua Bengio ("Learning Long-Term Dependencies with Gradient Descent is Difficult"). The expression "vanishing gradient" refers to the fact that the computations for each new layer become less and less clear. It is a problem similar to calculating the probability of a chain of events: if you multiply a probability between 0 and 1 by another probability between 0 and 1 many times over, the result is always zero, even in the case in which all those numbers expressed probabilities of 99%. A network with many layers is difficult to train because the "weights" of the last layer end up being too weak.

In 1997 Sepp Hochreiter and his professor Jurgen Schmidhuber came up with a solution: the Long Short Term Memory (LSTM) model. In this model, the unit of the neural network (the "neuron") is replaced by one or more memory cells. Each cell functions like a mini-Turing machine, performing simple operations of read, write, store and erase that are triggered by simple events. The big difference with Turing machines is that these are not binary decisions but "analog" decisions, represented by real numbers between 0 and 1, not just 0 and 1. For example, if the network is analyzing a text, a unit can store the information contained in a paragraph and apply this information to a subsequent paragraph. The reasoning behind the LSTM model is that a recurrent neural network contains two kinds of memory: there is a short-term memory about recent activity, and there is a long-term memory which is the traditional "weights" of the connections that change based on this recent activity. The weights change very slowly as the network is being trained. The LSTM model tries to retain information contained in the recent activity that traditional networks only used to fine-tune the weights before discarding them.

For 60 years it was assumed that no computing device can be more powerful than a Universal Turing Machine. Hava Siegelmann proved mathematically that analog RNNs can achieve super-Turing computing ("On the Computational Power of Neural Nets", 1992). Alan Turing himself had tried to imagine a way to extend the computational power of his universal machine ("Systems of Logic Based on Ordinals", 1938), but his idea cannot be implemented in practice. Siegelmann's system was not the first system to break the Turing limit using real numbers, and nobody has built a computer yet that can perform operations on real numbers in a single step.

Teaser: Machine Ethics

If we ever create a machine that is a fully-functioning brain totally equivalent to a human brain, will it be ethical to experiment on it? Will it be ethical to program it? Will it be ethical to modify it, and to destroy it at the end?

How not to Build an Artificial General Intelligence – Part 1: The Many-task Mind

In April 2013 i saw a presentation at Stanford's Artificial Intelligence Lab by the team of Kenneth Salisbury in collaboration with Willow Garage about a robot that can take the elevator and walk upstairs to buy a cup of coffee. This implies operations that are trivial for humans: recognizing that a transparent glass door is a door (not just a hole in the wall and never mind the reflection of the robot itself in the glass), identifying the right type of door (revolving, sliding or automatic), finding the handle to open the door, realizing that it's a spring-loaded door so it doesn't open as easily as regular doors, finding the elevator door, pressing the button to call the elevator, entering the elevator, finding the buttons for the floors inside an elevator whose walls are reflective glass (therefore the robot keeps seeing reflections of itself), pressing the button to go upstairs, locating the counter where to place the order, paying, picking up the coffee, and all the time dealing with humans (people coming out of the door, sharing the space in the elevator, waiting in line) and avoiding unpredictable obstacles; if instructions are posted, read the instructions, understand what they mean (e.g. this elevator is out of order or the coffee shop is closed) and change plan accordingly. Eventually, the robot got it right. It took the robot 40 minutes to return with the cup of coffee. It is not impossible. It is certainly coming. I'll let the experts estimate how many years it will take to have a robot that can go and buy a cup of coffee upstairs in all circumstances (not just those programmed by the engineer) and do it in 5 minutes like humans do. The fundamental question, however, is whether this robot can be considered an intelligent being because it can go and buy a cup of coffee or it is simply another kind of appliance.

It will take time (probably much longer than the optimists claim) but some kind of "artificial intelligence" is indeed coming. How soon depends on your definition of artificial intelligence. One of the last things that John McCarthy wrote before dying was: "We cannot yet characterize in general what kinds of computational procedures we want to call intelligent" (2007).

Nick Bostrom wrote that the reason A.I. scientists have failed so badly in predicting the future of their own field is that the technical difficulties have been greater than they expected. I don't think so. I think those scientists had a good understanding of what they were trying to build. The reason why "the expected arrival date [of artificial intelligence] has been receding at a rate of one year per year" (Bostrom's estimate) is that we keep changing the definition. There never was a proper definition of what we mean by "artificial intelligence" and there still isn't. No wonder that the original A.I. scientists were not concerned with safety or ethical concerns: of course, the machines that they had in mind were chess players and theorem provers. That's what "artificial intelligence" originally meant.

Being poor philosophers and poor historians, they did not realize that they belonged to the centuries-old history of automation, leading to greater and greater automata. And they couldn't foresee that, within a few decades, all these automata would become millions of times faster, billions of times cheaper, and would be massively interconnected. The real progress has not been in A.I. but in miniaturization. Miniaturization has made it possible to use thousands of tiny cheap processors and to connect them massively. The resulting "intelligence" is still rather poor, but its consequences are much more intimidating.

To start with, it is wise to make a distinction between an artificial intelligence and an A.G.I. (artificial general intelligence). Artificial intelligence is coming very soon if you don't make a big deal of it, and it might already be here: we are just using a quasi-religious term for "automation", a process that started with the waterwheels of ancient Greece if not earlier. Search engines (using very old fashioned algorithms and a huge number of very modern computers housed in "server farms") will find an answer to any question you may have. Robots (thanks to progress in manufacturing and to rapidly declining prices) will become pervasive in all fields, and become household items, just like washing machines and toilets; and eventually some robots will become multifunctional (just like today's smartphones combine the functions of yesterday's watches, cameras, phones, etc; and, even before smartphones, cars acquired a radio and an air conditioning unit, and planes acquired all sorts of sophisticated instruments).

Millions of jobs will be created to take care of the infrastructure required to build robots, and to build robots that build robots, required to build robots, and to build robots that build robots, and ditto for search engines, websites and whatever comes next. Some robots will come sooner, some will take centuries. And miniaturization will make them smaller and smaller, cheaper and cheaper. At some point we will be surrounded for real by Neil Stephenson's "intelligent dust" (see his novel "Diamond Age"), i.e. by countless tiny robots each performing one function that used to be exclusive to humans. If you want to call these one-function programs "artificial intelligence", suit yourself.

We wouldn't call "intelligent" a human being whose brain can do only one thing.

An AGI, instead, would be more like us: maybe none of us does anything well, but we do many things and we are capable of doing many things that we will actually never do. An AGI would not be limited to one or two or twenty tasks: it would be able to perform ALL the tasks that human beings perform, although not necessarily excel at any of them.

Making predictions about the coming of an AGI without having a clear definition of what constitutes an AGI is as scientific as making predictions about the coming of Jesus. An AGI could be implemented as a collection of one-function programs, each one specialized in performing one specific task. In this case someone has to tell the A.I. specialist which tasks we expect from an AGI. Someone has to list whether AGI requires being able to ride a bus in Zambia and to exchange money in Haiti or whether it only requires the ability to sort out huge amounts of data at lightning speed or what else. Once we have that list, we can ask the world's specialists to make reasonable estimates and predictions on how long it will take to achieve each of the functions that constitutes the AGI.

This is an old debate. Many decades ago the founders of computational mathematics (Alan Turing, Claude Shannon, Norbert Wiener, John von Neumann and so forth) discussed which tasks can become "mechanic", i.e. performed by a computing machine, i.e. can and cannot be computed, i.e. can be outsourced to a machine and what kind of machine it has to be. Today's computers that perform today's deep-learning algorithms, such as playing go/weichi, are still Universal Turing Machines, subject to the theorems proven for those classes of machines. Therefore, Alan Turing's original work still applies. The whole point of inventing (conceptually) the Turing Machine in 1936 was to prove whether a general algorithm to solve the "halting problem" for all possible program-input pairs exists, and the answer was a resounding "no": there is always at least one program that cannot be "decided", i.e. will never halt. And in 1951 Henry Gordon Rice generalized this conclusion with an even more formidable statement, "Rice's Theorem": any nontrivial property about the behavior of a Turing machine is undecidable, a much more general statement about the undecidability of Turing machines. In other words, it is proven that there is a limit to what machines can "understand", no matter how much progress is made, if they are Universal Turing Machines (as virtually all of today's computers are).

Nonetheless, by employing thousands of these machines, the "brute force" approach has achieved sensational feats such as machines that can beat go/weichi champions and recognize cats. So you might be tempted to accept that an AGI will be created by sheer "brute force": creating a one-function program for each possible task and then somehow putting them all together in one machine that will then be able to carry out any human function.

Some of us doubt that the human mind works that way. We have seen no neurological evidence that the human brain is a collection of one-function programs. We have seen evidence of the opposite: that the human mind is capable of applying the skills of one function to a different function, and sometimes without even being told to do so. We are AGIs because our brain can approach new tasks and find a way to perform them even if nobody trained us to carry out such new tasks.

So the other way to build an AGI is to build a learning system that can transfer its skills and knowledge from one task to another as it acquires and refines them. This is an old A.I. program that harks back to at least 1991, when Satinder Singh of the University of Massachusetts published "Transfer of Learning by Composing Solutions of Elemental Sequential Tasks" and Lorien Pratt of the Colorado School of Mines published "Direct Transfer Of Learned Information Among Neural Networks". Tom Mitchell's group at Carnegie Mellon University became the world's center of excellence in transfer learning and multitask learning, as documented by

Sebastian Thrun's " Is Learning the N-Th Thing Any Easier Than Learning the First?" (1996) and Rich Caruana's "Multitask Learning" (1997). But not much has improved since Sebastian Thrun and Lorien Pratt curated the book "Learning to Learn" (1998).

How not to Build an Artificial General Intelligence - Part II: Smart, not Deep, Learning

Simply telling me that Artificial Intelligence and robotics research will keep producing better and smarter devices (that are fundamentally not "intelligent" the way humans are) does not tell me much about the chances of a breakthrough towards a different kind of machine that will match (general) human intelligence.

I don't know what such a breakthrough should look like, but i know what it doesn't look like. The machine that beat the world champion of go/weichi was programmed with knowledge of virtually every major go/weichi game ever played, and it was allowed to run millions of logical steps before making any move. That obviously put the human contender at a huge disadvantage. Even the greatest go/weichi champion with the best memory can only remember so many games. The human player relies on intuition and creativity, whereas the machine relies on massive doses of knowledge and processing. Shrink the knowledge base that the machine is using to the knowledge base that we have and limit the number of logical steps it can perform to the number of logical steps that the human mind can perform before it is timed out, and then we'll test how often it wins against ordinary players, let alone world champions.

Having a computer (or, better, a huge knowledge base) play chess against a human being is like having a gorilla fight a boxing match with me: i'm not sure what conclusion you could draw from the result of the boxing match about our respective degrees of intelligence.

I wrote that little progress has been made in Natural Language Processing. The key word is "natural". Machines can actually speak quite well in unnatural language, a language that is grammatically correct but from which all creativity has been removed: "subject verb object - subject verb object - subject verb object - etc." The catch is that humans don't do that. If i ask you ten times to describe a scene, you will use different words each time.

Language is an art. That is the problem. How many machines do we have that can create art? How far are we from having a computer that switches itself on in the middle of the night and writes a poem or draws a picture just because the inspiration came? Human minds are unpredictable. And not only adult human minds: pets often surprise us, and children surprise us all the time. When is the last time that a machine surprised you? (Other than surprising you because they are still so dumb). Machines simply do their job, over and over again, with absolutely no imagination.

Here is what would constitute a real breakthrough: a machine that has only a limited knowledge of all the go/weichi games ever played and is allowed to run only so many logical steps before making a move and that can still play well. That machine will have to use intuition and creativity. That's a machine that would probably wake up in the middle of the night and write a poem. That's a machine that would probably learn a human language in a few months just like even the most disadvantaged children do. That is a machine that would not translate "'Thou' is an ancient English word" into "'Tu' e` un'antica parola Inglese", and that will not stop at a red traffic light if it creates a dangerous situation.

I suspect that this will require some major redesigning of the very architecture of today's computers. For example, a breakthrough could be a transition from digital architectures to analog architectures. Another breakthrough could be a transition from silicon (never used by Nature to construct intelligent beings) to carbon (the stuff of which all natural brains are made). And another one, of course, could be the creation of an artificial being that is self-conscious.

Today it is commonplace to argue that in the 1970s A.I. scientists gave up too quickly on neural networks and connectionism. My gut feeling is that in the 2000s we gave up a bit too quickly on the symbolic processing (knowledge-based) program. Basically, we did to the logical approach what we had done before to the connectionist approach: in the 1970s neural networks fell into oblivion because knowledge-based systems were delivering practical results... only to find out that knowledge-based systems were very limited and that neural networks were capable of doing more.

My guess is that there was nothing wrong with the knowledge-based approach. Unfortunately, we never figured out an adequate way to represent human knowledge. Representation is one of the oldest problems in philosophy, and I don't think we got any closer to solving it now that we have powerful computers. The speed of the computer does little to fix a wrong theory of representation.

So we decided that the knowledge-based approach was wrong and we opted for neural networks (deep learning and the likes). And neural networks have proven very good at simulating specialized tasks: each neural network does one thing well, but doesn't do what every human, even the dumbest one, and even animals, do well: use the exact same brain to carry out thousands (potentially an infinite number) of different tasks.

"If a machine is expected to be infallible, it cannot also be intelligent" (Alan Turing, 1947).

The Timeframe of Artificial General Intelligence

If by "artificial intelligence" we simply mean a machine that can do something (not everything) that we can do (like recognizing cats or playing chess), but not "everything" that we can do (both the mouse and the chess player do a lot of other things), then all machines and certainly all appliances qualify. Some of them (radio, telephone, television) are even forms of superhuman intelligence because they can do things that human brains cannot do.

Definitions do matter: there is no single answer to the questions "when will machines become intelligent" and "when will superhuman intelligence appear". It depends on what we mean by those words. My answer can be "it's already here" or "never".

As it stands, predictions about the future of (really) intelligent machines (of AGI) are predictions about a technology that doesn't exist. You can ask a rocket scientist for a prediction for when a human being will travel to Pluto: that technology exists and one can speculate what it will take to use that technology for that specific mission. On the contrary, my sense is that, using current technology, there is no way that we can create a machine that is even remotely capable of performing our routine cognitive tasks. The technology that is required does not yet exist. The machine that is supposed to become more intelligent than us and not only steal your job but even rule the world (and either kill us all or make us immortal) is pure imagination, just like angels and ghosts.

It is difficult to predict the future because we tend to predict one future instead of predicting all possible futures. Nobody (as far as i know) predicted that the idea of expert systems would become irrelevant in most fields because millions of volunteers would post knowledge for free on something called World-wide Web accessible by anybody equipped with a small computer-telephone. That was one possible future but there were so many possible futures that nobody predicted this one. By the same token, it is hard to predict what will make sense in ten years, let alone in 50 years.

What if 3D printing and some other technology makes it possible for ordinary people to create cheap gadgets that solve all sorts of problems. Why would we still need robots? What if synthetic biology starts creating alternative forms of life capable of all sorts of amazing functions. Why would we still need machines? There is one obvious future, the one based on what is around today, in which machines would continue to multiply and improve. There are many other futures in which computers and robots would become irrelevant because of something that does not exist today.

Anders Sandberg and Nick Bostrom, authors of "Whole Brain Emulation" (2008), conducted a "Machine Intelligence Survey" (2011) that starts with a definition of what an artificial intelligence should be: a system "that can substitute for humans in virtually all cognitive tasks, including those requiring scientific creativity, common sense, and social skills." My estimate for the advent of such a being is roughly 200,000 years: the timescale of natural evolution to produce a new species that will be at least as intelligent as us. If Artificial Intelligence has to be achieved by incremental engineering steps starting from the machines that we have today, my estimate about when a machine will be able to carry out a conversation like this one with you is: "Never". I am simply projecting the progress that i have witnessed in Artificial Intelligence (very little and very slow) and therefore i obtain an infinite time required for humans to invent such a machine.

But then, again, we'd probably have a lengthy discussion about what the expression "all cognitive tasks" really means. For example, leaving out consciousness from the category of cognitive tasks is like leaving out Beethoven from the category of musicians simply because we can't explain his talent.

As i wrote, machines are making us somewhat dumber (or, better the environments we design for automation make us dumber), and there is an increasing number of fields (from arithmetic to navigation) in which machines are now "smarter" than humans not only because machines got smarter but also because humans have lost skills that they used to have. If i project this trend to the future, there is a serious chance that humans will get so much dumber that the bar for artificial general intelligence will be lower and therefore rendering artificial intelligence more feasible than it is today; and "superhuman" intelligence may then happen, but it should really be called "subhuman" intelligence.

How NOT to Find a Breakthrough

Don't ask me what the breakthrough will be in A.I. If i knew it, i wouldn't be wasting my time writing articles like this one. But i have a hunch it has to do with recursive mechanisms for endlessly remodeling internal states: not data storage, but real "memory".

For historians a more interesting question is what conditions may foster such a breakthrough. In my opinion, it is not the abundance of a resource (such as computing power or information) that triggers a major paradigm shift but the scarcity of a resource. For example, James Watt invented the modern steam engine when and because Britain was in the middle of a fuel crisis (caused by the utter deforestation of the country). For example, Edwin Drake discovered petroleum ("oil") in Pennsylvania when and because whale oil for lamps was becoming scarce. Both innovations caused an economic and social revolution (a kind of "exponential progress") that completely changed the face of the world. The steam engine created an economic boom, reshaped the landscape, revolutionized transportation, and dramatically improved living conditions. Petroleum went on to provide much more than lighting to the point that the contemporary world is (alas) addicted to it. I doubt that either revolution would have happened in a world with infinite amounts of wood and infinite amounts of whale oil.

The very fact that computational power is becoming an infinite inexpensive resource makes me doubt that it will lead to a breakthrough in Artificial Intelligence.

Water power was widely available to Romans and Chinese, and they had the scientific know-how to create machines propelled by water; but the industrial revolution had to wait more than one thousand years. One reason (not the only one but a key one) why the Romans and the Chinese never started an industrial revolution is simple: they had plentiful cheap labor (the Romans had slaves, the Chinese emperors specialized in mobilizing masses of subjects).

Abundance of a resource is the greatest deterrent to finding an alternative to that resource. If "necessity is the mother of invention", as Plato said, then abundance is the killer of invention.

We live in the age of plentiful computational power. To some observers this looks like evidence that super-human machine intelligence is around the corner; to me this looks like evidence that our age doesn't even have the motivation to try.

My fear, in other words, is that the current success in "brute-force A.I." is slowing down (not accelerating) research in higher-level intelligence (the real meaning of "human intelligence"). If a robot can fix a car without knowing anything about cars, why bother to teach the robot how cars work? The success in (occasionally) recognizing cats, beating go/weichi champions and so forth is indirectly reducing the motivation to understand how the human mind (or, for that matter, the chimp's mind, or even a worm's mind) manages to recognize so many things in a split second and perform all sorts of actions. The success in building robots that perform this or that task with amazing dexterity is indirectly reducing the motivation to understand how the human mind can control the human body in such sophisticated manners in all sorts of situations and sometimes in completely novel ways.

Bill Joy wrote that "The future doesn't need us", but maybe it's the other way around: we won't need the future if the present starts giving us machines that can do everything we need.

The Real Breakthrough: Synthetic Biology

On the other hand, i have seen astonishing (quasi exponential) progress in Biotechnology, and therefore my estimate for when Biotech will be able to create an "artificial intelligence" is very different: it could already happen in one year. And my estimate of when Biotech might create a "superhuman" intelligence is also more optimistic: it could happen in a decade. I am simply basing my estimates on the progress that i have witnessed over the last 50 years; which might be misleading (again, most technologies eventually reach a plateau and then progress slows down), but at least this one has truly been "accelerating" progress. It would be interesting to discuss how Biotech might achieve this feat: will it be a new being created in the laboratory, or the intended or the accidental evolution of a being, or a cell by cell replica of the human body? But that's for another book.

The real deal is the digital to biological conversion that will increasingly allow biologists to create forms of life. That is indeed a "breakthrough". My guess is that machines will remain a tool (that every generation will brand "intelligent" and every generation will expect to get more "intelligent") but one of their applications, the biotech application, is likely to have the biggest impact on the future of life on this planet.

Biotechnology might even be better than A.I. for simulating and learning how the human brain works. For example, Madeline Lancaster at Cambridge University is using pluripotent human cells to grow three-dimensional tissues ("cerebral organoids") that she uses to model how the human brain develops ("Cerebral Organoids Model Human Brain Development And Microcephaly", 2013).

Humorous Intermezzo: Bayes and The End of the World

In 1983 physicist Brandon Carter introduced the "Doomsday Argument" later popularized by philosopher John Leslie in his book "The End of the World: The Science and Ethics of Human Extinction" (1996). This was a simple mathematical theorem based on Bayes's theorem demonstrating that we can be 95% certain that we are among the last 95% of all the humans ever to be born. Leslie calculated that we will reach this point in about 10,000 years. It has been tweaked up and down by various dissenters but, unlike Singularity Science, it sits on solid mathematical foundations: in fact, on the exact same foundations (Bayesian reasoning) as those buttressing today's deep-learning neural-networks like AlphaGo.

The Future of Miniaturization: the Next big Breakthrough?

If i am right and the widely advertised progress in machine intelligence is mainly due to rapid progress in miniaturization and cost reduction, then it would be more interesting to focus on the future of miniaturization. Whatever miniaturization achieves next is likely to determine the "intelligence" of future machines.

While IBM's Watson was stealing the limelight with its ability to answer trivial questions, others at IBM were achieving impressive results in Nanotechnology. In 1989 Don Eigler's team at IBM's Almaden Research Center, using the scanning tunneling microscope built in 1981 by Gerd Binnig and Heinrich Rohrer, carried out a spectacular manipulation of atoms that resulted in the atoms forming the three letters "IBM". In 2012 Andreas Heinrich's team (in the same research center) stored one magnetic bit of data in 12 atoms of iron, and a byte of data in 96 atoms; and in 2013 that laboratory "released" a movie titled "A Boy and His Atom" made by moving individual atoms.

This ability to trap, move and position individual atoms using temperature, pressure and energy could potentially create a whole new genealogy of machines.

The Real Future of Computing

In 1988 Mark Weiser envisioned a future in which computers will be integrated into everyday objects ("ubiquitous computing") and these objects will be connected with each other. This became known as the "Internet of Things" after 1998 when two MIT experts in Radio-frequency identification (RFID), David Brock and Sanjay Sarma, figured out a way to track products through the supply chain with a "tag" linking to an online database.

The technology is already here: sensors and actuators have become so cheap that embedding them into ordinary objects will not significantly increase the price of the object. Secondly, there are enough wireless ways to pick up and broadcast data that it is just a matter of agreeing on some standards. Monitoring these data will represent the next big wave in software applications.

Facebook capitalized on the desire of people to keep track of their friends. People own many more "things" than friends, spend more time with things than with friends, and do a lot more with things than with friends. The equivalent of Facebook for "things" does not exist yet, but potentially it is an order of magnitude bigger.

At the same time, one has to be aware that the proliferation of digital control also means that we will live in a world under systematic and all-pervasive surveillance in which machines will keep a record of everything that has happened and that is happening. Machines indirectly become spies. In fact, your computer (desktop, laptop, notepad, smartphone) is already a sophisticated and highly accurate spy that records every move you make: what you read, what you buy, whom you talk to, where you travel, etc. All this information is on your hard disk and can easily be retrieved by forensic experts.

The focus of computer science is shifting towards collecting, channeling, analyzing and reacting to billions of data arriving from all directions. Luckily, the "Internet of Things" will be driven by highly structured data.

The data explosion is proceeding faster than the increase in processing speed: exploring data is becoming increasingly more difficult with traditional John von Neumann computer architectures that were designed for calculations.

If I am sceptical about the creation of an agent that will be an artificial general intelligence, i am very aware that we are rapidly creating a sort of global intelligence as we connect more and more software and this giant network produces all sorts of positive feedback loops. This network is already out of control and gets harder to control with each passing year.

A Brief History of Bionic Humans, Cyborgs and Neuroengineering

I suspect that the science fiction novels and movies about cyborgs are more realistic than the ones about robots and artificial intelligences.

The first electrical implant in an ear was the work of French surgeons Andre Djourno and Charles Eyries in 1957. Building upon their work, in 1961 William House invented the "cochlear implant", an electronic implant that sends signals from the ear directly to the auditory nerve (as opposed to hearing aids that simply amplify the sound in the ear).

Spanish-born neuroscientist Jose Delgado is credited with publishing the first paper on implanting electrodes into human brains: "Permanent Implantation of Multi-lead Electrodes in the Brain" (1952). In 1965 he famously managed to control a bull via a remote device, injecting fear at will into the bull's brain. He then published his dystopian vision in the book "Physical Control of the Mind - Toward a Psychocivilized Society" (1969). In 1969 he created the first bidirectional brain-machine-brain interface when he implanted devices in the brain of a monkey and then sent signals in response to the brain's activity.

In 2000 William Dobelle in Portugal developed an implanted vision system that allowed blind people to see outlines of the scene. His patients Jens Naumann and Cheri Robertson became "bionic" celebrities as Dobelle continued to refine his artificial vision system.

The electrical interfacing of semiconductors and neurons is not trivial because neurons communicate using ions whereas semiconductors use electrons. In 1991 Peter Fromherz at the Max Planck Institute in Munich solved the problem of sensing the electrical field of a neuron on an electronic chip, and in 1995 he solved the problem of stimulating a neuron with an electronic chip (he used neurons of leeches). In 2001 he was therefore able to build a hybrid circuit of electronics and neurons (a snail's neurons).

In 2002 John Chapin and Sanjiv Talwar at the State University of New York debuted their "roborats", rats whose brains were fed electrical signals via a remote computer to guide their movements.

As for getting data out of the brain into a machine (output neuroprosthetics) in 1998 the Irish-born scientist Philip Kennedy at Georgia Tech developed a brain implant that could capture the "will" of a paralyzed man (Johnny Ray) to move an arm. In 1987 Kennedy had founded Neural Signals to develop a brain-computer interface, the first bionic startup. (Ray died in 2002 and in 2014 Kennedy himself almost died when he courageously chose to have electrodes surgically implanted in his own brain)

In 1998 Kevin Warwick at the University of Reading in Britain implanted a transmitter in his arm to activate computer-controlled devices, a bionic precursor of the "Internet of Things". (In the same year Warwick also created an artificial intelligence to compose pop songs). In 2002 Warwick used a BrainGate device to connect his nervous system to the Internet.

In 2002 Brown University spun off Cyberkinetics, a startup charged with developing its BrainGate technology. In 2005 John Donoghue's team implanted a BrainGate device in the brain of a paralyzed woman, Cathy Hutchinson, which allowed her to operate a robotic arm.

In 2002 the Brazilian-born scientist Miguel Nicolelis at Duke University implanted a microchip into a monkey's brain that allowed the monkey to control a robotic arm.

In 2004 Theodore Berger at the University of Southern California in Los Angeles demonstrated a hippocampal prosthesis conceived to replace the long-term-memory function lost by a damaged hippocampus. His lab would become another major hub of bionic research. In 2011 Berger developed "memory chips" that can turn memories on and off in a mouse's brain, and in 2015 Berger and Dong Song built a brain prosthesis to help people suffering from memory loss

In 2004 color-blind artist Neil Harbisson (born in Britain, raised in Spain, relocated to New York) became the first person in the world to have an antenna implanted in his skull, a device that transformed color into sound.

In 2004 PositiveID in Florida started selling VeriChip, an RFID chip implant for humans developed in Texas at Destron Fearing, a company that manufactures RFID tags for animal identification.

This is when the government stepped in. In 2006 the Defense Advanced Research Projects Agency (DARPA) asked scientists to submit "innovative proposals to develop technology to create insect-cyborgs".

And this is also when the transhumanist movement adopted bionics. In 2006 Seattle-based transhumanist Amal Graafstra boasted a microchip in each hand, one for storing data (that could be uploaded and downloaded from/to a smartphone) and one for a code that unlocked his front door and logged him into his computer. In 2012 Graafstra implanted chips on attendees of the Toorcamp for $50 each, and in 2013 he started a website to sell home implants, dangerousthings.com.

In 2010 Epoc in Australia released a neuroheadset for videogames, Emotiv, to play videogames with your brain waves.

The laboratory of Finnish-born engineer Arto Nurmikko at Brown University that had inherited the BrainGate project from Cyberkinetics. By 2008 this device had become a wireless transmitter for paralyzed patients with a neural implant that bypassed the spinal cord. In 2011 Leigh Hochberg of that team used BrainGate to make a paralyzed woman operate a robotic arm simply by thinking about the movement.

Experiments on brains became more and more ambitious. In 2011 Matti Mintz in Israel replaced a rat's cerebellum with a computerized cerebellum. In 2012 the brain implant designed by Sam Deadwyler at Wake Forest University managed to improve the long-term memory of monkeys.

At the same time some independents began to view implants as the tattoos of the 21st century. In 2013 biohacker Rich Lee in Utah hired Steve Haworth in Arizona to implant headphones into his ears. Haworth had pioneered "body modification", a high-tech evolution of "body piercing" that implants devices (typically magnets) under the skin.

Two-way transmission was just a matter of combining existing technologies. In 2013 Nicolelis made two rats communicate (and they were located in two different countries) by capturing the "thoughts" of one rat's brain and sending them to the other rat's brain over the Internet and an electrode. In 2015 Nicolelis connected the brains of monkeys so that they could collaborate to perform a task.

In 2013 the Indian-born computer scientist Rajesh Rao and the Italian-born psychologist Andrea Stocco at the University of Washington devised a way to send a brain signal from Rao's brain to Stocco's hand over the Internet, i.e. Rao made Stocco's hand move, probably the first time that a human was capable of controlling the body part of another human. In that year Rao (also a scholar of the ancient Indus script and of classical Indian painting) published "Brain-Computer Interfacing" (2013).

It was just a matter of time before someone thought of expanding the cyborgs beyond vision, sound and movement. In 2014 the team led by Italian-born electrical engineer Silvestro Micera at the Federal Institute of Technology (EPFL) in Switzerland designed an artificial hand for an amputee, Dennis Aabo-Soerensen. This hand sends electrical signals to the nervous system so as to create the sensation of touch.

In 2014 Chinese-born wireless scientist Ada Poon at Stanford invented a safe way to transfer energy to chips implanted in the body (to "electroceutical devices").

In 2015 Zoran Nenadic and An Do of the University of California at Irvine attached an electroencephalograph device to the head of a paraplegyc man and made him walk a few steps.

In 2015 EPFL built a robotic wheelchair for paralyzed people. This chair combines brain control with artificial intelligence. In 2016 Gregoire Courtine at EPFL used BrainGate to restore movement to a monkey's paralyzed leg. In 2016 Nick Ramsey's team in Holland (at University Medical Center Utrecht) inserted wireless electrodes into the skull of a paralyzed patient (unable to speak or move) so that she could control a computer mouse simply by thinking of moving her fingers.

After Machine Inteligence: Machine Creativity – Can Machines do Art?

The question "Can machines think?" is rapidly becoming obsolete. I have no way of knowing whether you "think". We cannot enter someone else's brain and find out if that person has feelings, emotions, thoughts, etc. All we know about other people's inner lives is that it generates a behavior very similar to our own, and therefore we conclude that other people too must have the same kind of inner lives that we have (feelings, emotions, thoughts, etc). Since we cannot even determine with absolute certainty the consciousness of other people, it sounds a bit useless to discuss whether machines can be conscious. Can machines think? Maybe, but we'll never find out for sure, just like we'll never find out for sure if all humans think.

The question "Can machines be creative?" is much more interesting. Humans have always thought of themselves as creative beings, but always failed to explain what that really means. The humble spider can make a very beautiful spider-web. Some birds create spectacular nests. Bees perform intricate dances. Most humans don't think that the individual spider or the individual bird is a "creative being". Humans assume that something in its genes made it do what it did, no matter how complex and brilliant. But what exactly is different between the spider or the bird and Shakespeare, Michelangelo or Beethoven?

Humans use tools to make art (if nothing else, a pen). But the border between artist and tool has gotten blurred since Harold Cohen conceived AARON, a painting machine, in 1973. Cohen asked: "What are the minimum conditions under which a set of marks functions as an image?" I would rephrase it as "What are the minimum conditions under which a set of signs functions as art?" Even Marcel Duchamp's "Fountain" (1917), which is simply a urinal, is considered "art" by the majority of art critics. Abstract art is mostly about... abstract signs. Why are Piet Mondrian's or Wassily Kandinsky's simple lines considered art? Most paintings by Vincent Van Gogh and Pablo Picasso are just "wrong" representations of the subject: why are they art, and, in fact, great art?

During the 1990s and 2000s several experiments further blurred that line: Ken Goldberg's painting machine "Power and Water" at the University of South California (1992); Matthew Stein's

PumaPaint at Wilkes University (1998), an online robot that allows Internet users to create original artwork; Jurg Lehni's graffiti-spraying machine Hektor in Switzerland (2002); the painting robots developed since 2006 by Washington-based software engineer Pindar Van Arman; and Vangobot (2008) (pronounced "Van Gogh bot"), a robot built by Nebraska-based artists Luke Kelly and Doug Marx that renders images according to preprogrammed artistic styles. After a Kickstarter campaign in 2010, Chicago-based artist Harvey Moon built drawing machines, set their "aesthetic" rules, and let them do the actual drawing. In 2013 Oliver Deussen's team at the University of Konstanz in Germany demonstrated e-David (Drawing Apparatus for Vivid Interactive Display), a robot capable of painting with real colors on a real canvas. In 2013 the Galerie Oberkampf in Paris showed paintings produced over a number of years by a computer program, "The Painting Fool", designed by Simon Colton at Goldsmiths College in London. The Living Machines exhibition of 2013 at London's Natural History Museum and Science Museum featured "Paul", a creative robot capable of sketching a portrait, developed by French inventor Patrick Tresset since 2011, and BNJMN (pronounced "Benjamin"), a robot capable of generating images built for the occasion by Travis Purrington and Danilo Wanner from the Basel Academy of Art and Design.

While each of these systems caused headlines in the press, none was autonomous and the "trick" was easy to detect.

Then deep learning happened. Deep learning consists in a multi-layer network that is trained to recognize an object. The training consists in showing the network many instances of that object (say, many cats). Andrew Zisserman's team at Oxford University was probably the first to think of asking a neural network to show what it was learning during this training ("Deep Inside Convolutional Networks", 2014). Basically, they used the neural network to generate the image of the object being learned (say, what the neural network has learned a cat to be like).

In May 2015 a Russian engineer at Google's Swiss labs, Alexander Mordvintsev, used that idea to make a neural network produce psychedelic images. One month later he posted a paper titled "Inceptionism" (jointly with Christopher Olah, an intern at Jeff Dean's Google Brain team in Silicon Valley, and with Mike Tyka, an artist working for Google in Seattle) that sort of coined a new art movement. Neural nets trained to recognize images can be run in reverse so that they instead generate images. More importantly, the networks can be asked to identify objects that actually don't exist, like when you see a face in a cloud. By feeding back this "optical illusion" into the network over and over again, the network eventually displays a detailed image, which is basically the machine's equivalent of a human hallucination. For example, a neural network trained to recognize animals will identify inexistent animals in a cloudy sky.

In August 2015 two students (Leon Gatys and Alexander Ecker) of Matthias Bethge's lab at the University of Tubingen in Germany submitted a paper titled "A Neural Algorithm of Artistic Style" in which they showed that neural networks can be used to imitate the style of any Maestro. A neural network trained to recognize an object tends to separate content and style, and the "style" side of it can be applied to other objects, therefore obtaining a version of those objects in the style that the network previously learned.

In September 2015, at the International Computer Music Conference, Donya Quick, a composer working at Paul Hudak's lab at Yale University, presented a computer program called Kulitta for automated music composition. In February 2016 she published on Soundcloud a playlist of Kulitta-made pieces.

In February 2016 Google staged an auction of 29 paintings made by its artificial intelligence at the Grand Theater in San Francisco in collaboration with the Gray Area Foundation for the Arts ("DeepDream: The Art of Neural Networks").

In March 2016 a 20-year-old Princeton University student, Ji-Sung Kim, and his friend Evan Chow created a neural network that can improvise like a jazz musician on Pat Metheny's "And Then I Knew" (1995).

In April 2016 a new Rembrandt portrait was unveiled in Amsterdam, 347 years after the painter's death: Joris Dik at Delft University of Technology created this 3D-printed fake Rembrandt consisting of more than 148 million pixels based on 168,263 fragments from 346 of Rembrandt's paintings. (To be fair, a similar feat had been achieved in 2014 by Jeroen van der Most whose computer program had generated a "lost Van Gogh" after analyzing statistically 129 real paintings of the master).

In May 2016 Daniel Rockmore at Dartmouth College organized the first Neukom Institute Prizes in Computational Arts (soon nicknamed the "Turing Tests in the Creative Arts"), that included three contests to build computer programs that can create respectively a short story, a sonnet, and a DJ set. Spanish students Jaume Parera and Pritish Chandna won the prize for the DJ set, while three students of Kevin Knight's lab at the University of Southern California won the prize for the sonnet ("And from the other side of my apartment/ An empty room behind the inner wall/ A thousand pictures on the kitchen floor/ Talked about a hundred years or more").

In July 2016 a Bay Area software engineer, Karmel Allison, launched CuratedAI, an online magazine of poems and prose written by A.I. programs.

In May 2016 the TED crowd got to hear a talk by Blaise Aguera y Arcas, principal scientist at Google, titled "We're on the edge of a new frontier in art and creativity — and it's not human".

The standard objection to machine art is that the artwork was not produced by the machine: a human being designed the machine and programmed it to do what it did, hence the machine should get no credit for its "artwork". Because of their nonlinearity, neural networks distance the programmer from the working of the program, but ultimately the same objection holds.

However, if you are painting, it means that a complex architecture of neural processes in your brain made you paint, and those processes are a result of the joint work of a genetic program and of environmental forces. Why should you get credit for your artwork?

If what a human brain does is art, then what a machine does is also art.

A sceptical friend, who is a distinguished art scholar at UC Berkeley, told me: "I haven't seen anything I'd take seriously as art". But that's a weak argument: many people cannot take seriously as art the objects exhibited in museums of contemporary art, not to mention performance art, body art and dissonant music. How does humankind decide what qualifies as art?

The Turing Test of art is simple. We are biased when they tell us "this was done by a computer". But what if they show us the art piece and tell us it was done by an Indonesian artist named Namur Saldakan? I bet there will be at least one influential art critic ready to write a lengthy analysis of how Saldakan's art reflects the traditions of Indonesia in the context of globalization etc., etc.

In fact, the way that a neural network can be "hijacked" to do art may help us to understand the brain of the artist. It could lead to a conceptual breakthrough for neuroscientists. After all, nobody ever came up with a decent scientific theory of creativity. Maybe those who thought of playing the neural net in reverse told us something important about what "creativity" is.

This machine art poses other interesting questions for the art world.

What did the art collectors buy at the Google auction? The output of a neural network is a digital file, which can be copied in a split second: why would you pay for something, of which an unlimited number of copies can be made? In order to guarantee that no other copies will ever be made, we need to physically destroy the machine or... to re-train the neural network so it will never generate those images again.

Who appreciates human art? Humans. We even have professionals called "art critics" who spend their entire life doing just that. Who appreciates machine art? The same humans. That is where the notion of art diverges. Human art is for humans. It will influence humans. It is part of human history. Faced with machine art, we try to fit machine art into the human narrative. This introduces an asymmetry between human art and machine art. To have full symmetry, it is not enough to have a machine that produces art. You also need machines that can appreciate that art and that can place it in a historical and social context; otherwise it is still missing something that human art has.

Machine art shows that it is not difficult to be creative, but it is difficult to be creative in a way that matters.

The Moral Issue: Who's Responsible for a Machine's Action?

During the 2000s, drones and robotic warfare stepped out of science-fiction movies and into reality. According to the Bureau of Investigative Journalism, an independent non-profit organization founded by David and Elaine Potter in 2010, US drones have killed between 2500 and 4,000 people in at least seven countries (Afghanistan, Pakistan, Syria, Iraq, Yemen, Libya and Somalia). About 1,000 of them were civilians, about 200 were children.

These weapons represent the ultimate example of how machines can relieve us of the sense of guilt. If i accidentally kill three children, i will feel guilty for the rest of my life and perhaps commit suicide. But who feels guilty if the three children are killed by mistake by a drone that was programmed 5.000 kms away by a team using Google maps, Pakistani information and Artificial Intelligence software, a strike authorized by a general or by the president in person? The beauty of delegating tasks to machines is that we decouple the action from the perpetrator. We dilute the responsibility so much that it becomes easier to "pull the trigger" than not to pull it. What if the mistake was due to malfunctioning software? Will the software engineer feel guilty? She may not even learn that there was a "bug" in her piece of software; and, if she does, she may never realize that the bug caused the death of three children.

This process of divorcing the killing from the killer is not new. It started at least in first World War with the first aerial bombings (a practice later immortalized by Pablo Picasso, when it still sounded horrible, in his painting "Guernica") and that happened precisely because humans were using machines (the airplanes) to drop the bombs on invisible citizens instead of throwing grenades or shooting guns against visible enemies. The killer will never know nor see the people he killed.

What applies to warfare applies to everything else. The use of machines to carry out an action basically relieves the machine's designers and its operators of real responsibility for that action.

The same concept can be applied, for example, to surgery: if the operation performed by a machine fails and the patient dies, who is to blame? The team that controlled the machine? The company that built the machine? The doctor who prescribed the use of that specific machine? I suspect that none of these will feel particularly guilty. There will simply be a counter that will mechanically add one to a statistical number of failed procedures. "Oops: you are dead". That will be society's reaction to a terrible incident.

You don't need to think of armed drones to visualize the problem. Think of a fast-food chain. You order at a counter, then you move down the counter to pay at the cash register, and then you hang out by the pick-up area. Eventually some other kid will bring you the food that you ordered. If what you get is not what you ordered, it is natural to complain with the kid who delivered it; but he does not feel guilty (correctly so) and his main concern is to continue his job of serving the other customers who are waiting for their food. In theory, you could go back to the ordering counter, but that would imply either standing in line again or upsetting the people who are in line. You could summon the manager, who was not even present when the incident happened, and blame him for the lousy service. The manager would certainly apologize (it is his job), but even the manager would be unable to pinpoint who is responsible for the mistake (the kid who took the order? the chef? the pen that wasn't writing properly?)

In fact, many businesses and government agencies neatly separate you from the chain of responsibility so that you will not be able to have an argument with a specific person. When something goes wrong and you get upset, each person will reply "I just did my job". You can blame the system in its totality, but in most cases nobody within that system is guilty or gets punished. And, still, you feel that the system let you down, that you are the victim of an unfair treatment.

This manner of decoupling the service from the servers has become so pervasive that younger generations take it for granted that often you won't get what you ordered.

The decoupling of action and responsibility via a machine is becoming pervasive now that ordinary people use machines all the time. Increasingly, people shift responsibility for their failures to the machines that they are using. For example, people who are late for an appointment routinely blame their gadgets. For example, "The navigator sent me to the wrong address" or "The online maps are confusing" or "My phone's batteries died". In all of these cases the implicit assumption is that you are not responsible, the machine is. The fact that you decided to use a navigator (instead of asking local people for directions) or that you decided to use those online maps (instead of the official government maps) or that you forgot to recharge your phone doesn't seem to matter anymore. It is taken for granted that your life depends on machines that are supposed to do the job for you and, if they don't, it is not your fault.

There are many other ethical issues that are not obvious. Being a writer who is bombarded with copyright issues all the time, here is one favorite. Let us imagine a future in which someone can create an exact replica of any person. The replica is just a machine, although it looks and feels and behaves exactly like the original person. You are a pretty girl and a man is obsessed with you. That man goes online and purchases a replica of you. The replica is delivered by mail. He opens the package, enters an activation code and the replica starts behaving exactly like you would. Nonetheless, the replica is, technically and legally speaking, just a toy. The manufacturer guarantees that this toy has no feelings/emotions, it simply simulates the behavior that your feelings/emotions would cause. Then this man proceeds to abuse that replica of you and later it "kills" it. This is a toy bought from a toy store, so it is perfectly legal to do anything the buyer wants to do with it, even to rape it and even to kill it. I think you get the point: we have laws that protect this very sentence that you are reading from being plagiarized and my statements from being distorted, but no law protects a full replica of us.

Back to our robots capable of critical missions: since they are becoming easier and cheaper, they are likely to be used more and more often to carry out these mission-critical tasks. Easy, cheap and efficient: no moral doubts, no second thoughts, no double crossing. The temptation to use machines instead of humans whenever the ethical boundaries are fuzzy will be too strong to resist.

The program of neural networks is increasingly becoming a program for building a silicon copy of the human brain, which will then pilot a body to perform human-level tasks. Hidden behind this program is the unspoken goal of human nature: deprive other human beings of their rights and make them work for us. The neural network that will achieve full parity with the human brain will, ultimately, be a human being without human rights. We can do anything we want to a machine and to the software that the machine is running, whereas we have laws that limit what we can do to other humans. When we have a machine that is fully equivalent to a human being, we will be able to satisfy our secret desire to use human beings without having to worry about their rights.

I wonder if it is technology that drives the process of de-responsabilization or it is the desire to be relieved of moral responsibility that drives the adoption of new technology. I wonder whether society is aiming for the technology that minimizes our responsibilities rather than aiming for the technology that maximizes our effectiveness. What society should do instead is aim for the technology that maximizes our accountability.

The Dangers of Machine Intelligence: Machine Credibility

The world has indeed changed: these days humans have more faith in machines than in gods.

GPS mapping and navigation software is not completely reliable when you drive on secondary mountain roads. When my hiking group is heading to the mountains, we have to turn on the most popular "navigator" because some of my friends insist on using it even if there is someone in the car who knows the route very well. They will stop the car if the navigation system stops working. And they tend to defend the service even when faced with overwhelming evidence that it took us to the wrong place or via a ridiculous route.

In September 2013 i posted on Facebook that YouTube was returning an ad about (sic) pooping girls when i looked for "Gandhi videos". An incredible number of people wrote back that the ad was based on my search history. I replied that i was not logged into YouTube, Gmail or any other product. A friend (who has been in the software industry all his life) then wrote "It doesn't matter, Google knows". It was pointless to try and explain that if you are not logged in, the software (whether Google, Bing or anything else) does not know who is doing the search (it could be a guest of mine using my computer, or it could be someone who just moved into my house using the same IP address that i used to have). And it was pointless to swear that i had never searched for pooping girls! (for the last week or so i had been doing a research to compile a timeline of modern India). Anyway, the point is not that i was innocent, but that an incredible number of people were adamant that the software knows that i am the one doing that search. People believe that the software knows everything that you do. It reminded me of the Catholic priest in elementary school: "God knows!"

Maybe we're going down the same path. People will believe that software can perform miracles when in fact most software has bugs that make it incredibly stupid.

Maybe we are witnessing what happened in ancient times with the birth of religions. (Next they started burning at the stakes the heretics like me who refused to believe).

The faith that an ordinary user places in a digital gadget wildly exceeds the faith that its very creators place in it.

If i make a mistake just once giving directions, i lose credibility for a long time; if the navigation system makes a mistake, most users will simply assume it was an occasional glitch and will keep trusting it. The tolerance for mistakes seems to be a lot higher when it comes to machines.

People tend to believe machines more than they believe humans, and, surprisingly, seem to trust machine-mediated opinions better than first-hand opinions from an expert. For example, they will trust the opinions expressed on websites like Amazon or Yelp more than trusting the opinion of the world's experts on books and restaurants. They believe their navigation system more than they believe someone who has spent her entire life in the neighborhood.

The evidence (e.g. political elections) show that we are a lot less smart than we think, and we can easily be fooled by humans. When we use a computer, we seem to become even more gullible. Think of how successful "spam" is, or even of how successful the ads posted by your favorite search engine and social media are. If we were smarter, those search engines and social media would rapidly go out of business. They thrive because millions of people click on those links.

The more "intelligent" software becomes, the more likely that people trust it. Unfortunately, at the same time the more "intelligent" it becomes, the more capable of harming people it will be. It doesn't have to be "intentionally" evil: it can just be a software bug, one of the many that software engineers routinely leave behind as they roll out new software releases that most of us never asked for.

Imagine a machine that broadcasts false news, for example that an epidemic is spreading around New York killing people at every corner. No matter what the most reputable reporters write, people will start fleeing New York. Panic would rapidly spread, from city to city, amplified by the very behavior of the millions of panicking citizens (and, presumably, by all the other machines that analyze, process and broadcast the data fed by that one machine).

In June 2016 Baidu published the news that an Indian woman gave birth to 11 twins. Those of you who are old enough will remember this story. It was false the first time it came out in 2011 and it is still false today, but it keeps being repeated on websites throughout the world. The Baidu spider simply scours the web for interesting news and has no way to find out whether the news is correct or not. An investigative report, or for that matter any intelligent being with 20 minutes to spare, can easily find out that the news was fabricated in 2011 (in Zambia, apparently). The scary thing is not that the spiders are dumb enough to believe all sorts of scams; the scary thing is that this becomes a news on Baidu, the main source of news in China. Millions of Chinese people are now convinced that (quote) "A Woman Gave Birth to 11 Babies at a Time in India".

Drone strikes seem to enjoy the tacit support of the majority of citizens in the USA. That tacit support arises not only from military calculations (that a drone strike reduces the need to deploy foot soldiers in dangerous places) but also from the belief that drone strikes are accurate and will mainly kill terrorists. However, drone strikes that the USA routinely hails as having killed terrorists are often reported by local media and eyewitnesses in Pakistan, Afghanistan, Yemen and so on as having killed a lot of harmless civilians, including children. People who believe that machines are intelligent are more likely to support drone strikes. Those who believe that machines are still very dumb are very unlikely to support drone strikes. The latter (including me) believe that the odds of killing innocents are colossal because machines are so dumb and are likely to make awful mistakes (just like the odds that the next release of your favorite operating system has a bug are almost 100%). If everybody were fully aware of how inaccurate these machines are, i doubt that drone programs would exist for much longer.

In other words, i am not so much afraid of machine intelligence as of human gullibility.

The Dangers of Machine Intelligence: Speed Limits for Machines?

In our immediate future i don't see the danger that future machines will be conceptually difficult to understand (superhuman intelligence), but i do see the danger that future machines will be so fast that controlling them will be a major project in itself. We already cannot control a machine that computes millions of times faster than our brain, and this speed will keep increasing in the foreseeable future.

That's not to say that we cannot understand what the machine does: we perfectly understand the algorithm that is being computed. In fact, we wrote it and fed it into the machine. It is computed at a much higher speed than the smartest mathematician could. When that algorithm leads to some automatic action (say, buying stocks on the stock market), the human being is left out of the loop and has to accept the result. When thousands of these algorithms (each perfectly understandable by humans) are run at incredible speed by thousands of machines interacting with each other, humans have to trust the computation. It's the speed that creates the "superhuman" intelligence: not an intelligence that we cannot understand, but an intelligence vastly "inferior" to ours that computes very quickly. The danger is that nobody can make sure that the algorithm was designed correctly, especially when it interacts with a multitude of algorithms.

The only thing that could be so fast is another algorithm. I suspect that this problem will be solved by introducing the equivalent of speed limits: algorithms will be allowed to compute at only a certain speed, and only the "cops" (the algorithms that stop algorithms from causing problems) will be allowed to run faster.

The Dangers of Machine Intelligence: Criminalizing Common Sense

There is something disturbing about the machines that intelligent humans are building with the specific mandate to overcome the individuality of intelligent humans. Stupid machines in charge of making sure that human intelligence does not interfere with rules and regulations are becoming widespread in every aspect of life.

I'll take a simple example because i find it even more telling than the ones that control lives at higher and more sinister levels. I live in the San Francisco Bay Area, one of the most technologically advanced regions in the world. We hold evening events in one of the most prestigious universities of the Bay Area. Because of a world-famous fog, the weather is chilly (if not cold) in the summer, especially in the evening. Nonetheless, the computers have been programmed to provide air conditioning throughout the campus for as long as there are people at work, no matter how cold it is getting outside. People literally bring sweaters and even winter coats at these evening classes. Never mind the total waste of energy; the interesting point is that nobody knows anymore how to tell the machines to stop doing that. After months of trying different offices, we still are "not sure who else to contact about it" (quoting the head of a department in the School of Science) "apparently it is very difficult to reset the building's thermostat".

This is the real danger of having machines run the world. I don't think any of us would call a thermostat "intelligent" when it directs very cold air into a room during a very cold evening. In fact, we view it as utterly stupid. However, it is very difficult for the wildly more intelligent race that created it to control its behavior according to common sense. The reason is that this incredibly stupid machine was created to overcome the common sense with which more intelligent beings are equipped. Think about it and probably your computer-controlled car, some of your computer-controlled appliances and systems around you often prohibit you from performing actions that common sense and ordinary intelligence would demand, even when those cars and appliances work perfectly well, and, in fact, precisely because they work perfectly well.

I am afraid of the millions of machines that will operate within human society with the specific goal of making sure that humans don't use common sense but simply follow the rules; in other words, with the specific goal of making us stupid.

The Dangers of Machine Intelligence: You Are a Budget

Another danger is that what will truly increase exponentially is the current trend to use computing power as a sales tool. The reason that people are willing to accept the terms and conditions of e-commerce websites is that these companies have been very good at concealing what they do with the information that they collect about you. The best minds of Hammerbacher's generation are thinking about how to make people click ads, and they do it by exploiting every tiny data that they can put their hands on. It's not only the best minds but also the best machines. Artificial Intelligence techniques are already being used to gather information on you (what used to be called "espionnage") for the purpose of targeting you with more effective sales strategies.

The original purpose of the World-wide Web was not to create a world in which smart software controls every move you make online and uses it to tailor your online experience; but that is precisely what it risks becoming. Computer science is becoming the discipline of turning your life into somebody else's business opportunity.

The Dangers of Machine Intelligence: Machine Stupidity

Perhaps the biggest danger of an artificial intelligence is its own stupidity. After the 2016 elections Google's answer to "who won the popular vote" was "Donald Trump", which was false at every stage of the vote counting process. The reason for that error is simple: a right-wing organization had easily found a way to become the top result on Google's search engine when that question was asked. You can easily fool an artificial intelligence, and this will remain true for many, many decades. Any service replaced by an artificial intelligence is (and will be) easily vulnerable to ill-intentioned people.

The Dangers of Machine Intelligence: Who will get there first?

Traditionally, most of the investment for A.I. research has come (directly or indirectly) from military agencies such as DARPA whose purpose is to build war machines. Now those big corporations whose business plan is based on advertising are also investing massively in A.I.

Judging from where the money is (or, better, comes from), one is tempted to conclude that the first general-purpose A.I., the first A.G.I., will emerge from either the warmongering military-industrial complex or from the greedy corporate world. We have to face the realistic scenario that the Singularity is more likely to arise as the descendant of a military machine or of an advertising algorithm than as the evolution of humanitarian software.

Dystopian Quartet 1. Andante: Lessons Learned from A.I.

The A.I. scientists of the 1950s have been mocked for making unreasonable predictions about the intelligence of computers. It turns out that they were mostly right. Those scientists lacked machines powerful enough to implement the theory, but, decades later, the theory can be implemented on powerful GPUs (especially when munificent benefactors, like Google, volunteer to buy thousands of them) and it works. Neural networks made of these powerful GPUs can indeed simulate a lot of "intelligent" tasks.

Therefore A.I. scientists have proven to neuroscientists that a lot of our "intelligence" is simply computational math, algorithms, formulas, very intricate systems of equations.

However, some tasks are too intelligent for us and still too intelligent also for the machine. For example, the original app: weather forecast. We still can't accurately predict the weather (except in sunny Silicon Valley). Weather forecast involves too many factors, and we don't have the dataset of past behavior that could help "recognize" the pattern for the future.

There is a second limitation to machine intelligence, and sometimes to human intelligence: lack of common sense. Common sense requires an almost infinite amount of knowledge about ordinary things. It turns out that the ordinary is much more difficult to map than the extraordinary. It is not difficult to describe mathematically how a master plays chess or weichi/go. It is very difficult to describe mathematically how a waiter cleans a table at the restaurant.

Common sense is hard to replicate with algorithms. It has proven much easier to remove the need for common sense. To a large extent, the history of human civilization is the history of removing the need for common sense. We don't need to run away from tigers anymore and we don't even need common sense to cross a very busy street. We have structured the environment so that life can be easy, safe, and predictable. In the "developed" world, there are rules and regulations for just about everything. We train children from a very early age to abandon common sense and follow rules. We strive to remove every possible glitch from our factories, subways, malls, and streets. Removing the need for common sense enables even the dumbest people (like me) to easily survive. I can write, teach, travel and enjoy my hobbies regardless of whether today my mind is sharp or not: there is a vast system of rules and regulations that takes care of me. We turned human behavior and the human environment into a well-organized, highly predictable, dumb, machine-friendly system. No wonder that machines can easily coexist with humans. The machines don't have to be very intelligent to deal with people because people have built an environment that doesn't require the one thing that machines don't have: common sense. And no wonder that we can easily replace humans with machines: humans have been trained all their life to behave like machines. So the issue now, as far as business is concerned, is not whether a machine can do the same job (it can, given a well-structured environment) but which one is overall cheaper to "operate".

We transferred "intelligence" from the human brain to the environment so that humans don't need intelligence anymore to survive in the environment; but that also means that machines can now do what only humans were capable of doing when the environment was chaotic and surviving in it required intelligence.

Dystopian Quartet 2. Allegro Forte: We are Surrounded

One day i walked into the subway station of a Chinese city and looked at the wall. It has machines that sell tickets, machines that sell drinks, machines that give you cash, machines that take passport-size photos of you and even machines that sell toys. I bought my ticket and then put my bag into a machine that detects metal and then inserted my ticket into a machine that validates the ticket. I walked into an escalator that took me to my train. And, of course, the train is a machine (controlled by machines).

MACHINES THAT SELL YOU TICKETS
MACHINES THAT SELL DRINKS
MACHINES THAT TAKE YOUR PICTURE
MACHINES THAT GIVE YOU CASH

We are already surrounded by machines. These machines perform a lot of jobs that used to be done by people, but today most people have better-paid jobs building or selling or maintaining these machines. The Chinese are wealthier today than when they didn't have machines.

(On top of being surrounded by machines, machines are also beginning to populate and colonize our body: a number of people already live in symbiosis with a machine that has been implanted or is attached to their body, whether an artificial heart or a hearing aid).

We do not worry that these boxed machines may someday conquer the world. We, instead, react emotionally when the machine looks like a human being and walks. This has more to do with psychology than with anything else. The machines that walk and look like humans are actually the least useful. Most of them are tourist attractions: they serve to attract people into the store. Buy a Pepper robot and the whole neighborhood will come to take a picture with the robot. These walking humanoids are far less useful than the immobile machines of the subway station.

Dystopian Quartet 3. A Cappella: We are being Programmed

The problem is not that we are surrounded by machines. The problem is that we are asked to behave like (very stupid) machines in order to make these machines useful. My recent flight from Beijing to San Francisco is typical. From the moment i entered the Beijing airport to the moment i exited the San Francisco airport i was behaving like a machine. I had to stand in one line after the other, and there were precise rules to follow at each line. In Beijing i even had to stand in a specific position so that the security guard's camera could take a good picture of me. In San Francisco i used an incredibly stupid terminal to get my passport checked. The whole experience requires that you abandon any notion of being a human being. This is the future, which is increasingly the present: we have to follow rules and regulations so that we can be surrounded by machines.

We are being programmed as much as the machines that surround us.

There is a moment at the airport when you don't have to follow rules and regulations: during the long journey from the security checkpoint to your gate. Before you reach your gate (no matter how late you are) they want you to shop at countless shops. Why? The money that you spend on your way to the gate helps them to pay for all those machines. That's the world we live in. Increasingly, we are asked to behave like machines (i.e. like stupid humans) when we interact with machines, so that machines can perform some useful task. We are given freedom to spend our money. That money is actually used to fund machines. Anything you do in your spare time has a cost, and that cost, ultimately, is money for machines. When you make a phone call to a friend, you are actually paying to support a chain of machines extending from your phone to your friend's phone via countless communication machines. When you watch a movie, you are paying to support machines that create, store, organize, and deliver entertainment. And so forth. You have no freedom when you interact with machines: you must obey precise rules and regulations just as if you were a machine. And during the few moments when you have freedom, you are actually spending money to pay for those machines. If your only purpose in society is to pay for the machines that surround you and to behave like a machine when you interact with those machines, the question "What is the meaning of life?" acquires a whole new dimension. Isn't it easier to replace you with machines even in your "free" moments? Why can't we just have machines building and controlling machines, and just remove you and your money from the loop? Are we slowly but steadily making humans redundant? What is the real meaning of this increasingly automated consumer society? Consuming so that we can pay for automation that helps us consume? Is the ultimate goal the "consuming" or the "automating"? What is left at the end? So far it looks like machines are expanding and humanity is retreating. Extrapolate, and you can foresee a future in which humans will be redundant, or, worse, just an annoyance.

Note that this is due to the stupidity (not intelligence) of machines. If machines were intelligent, we wouldn't need to lower our intelligence, to behave like machines. We have to create a highly structured and regulated society (i.e. to enforce machine-like behavior on humans) because otherwise machines would not be able to deal with us humans. The process of turning us into machines is being triggered not by the intelligence of machines but by their utter and probably incurable stupidity.

(The astute reader has already been asking "Who are they, the people who want to turn us into machines?" Of course, it's us. We are de facto committing collective suicide on a massive scale).

Dystopian Quartet 4. Adagio: Humanity without Humanity

We live in a partially automated world: we travel using machines (cars, buses, trains, planes), kitchen appliances do most of the household chores, machines such as television sets and personal computers provide our entertainment.

Our interaction with other humans is increasingly limited because machines perform many of the functions that used to be performed by humans. Who gives you cash at the bank? An automatic teller machine. Who hands you the ticket at the parking garage? A machine.

We tend to look at the machines that replace humans purely in economic terms: the service is now available 24/7 and it is cheap or even free; a job is lost; we can create more jobs elsewhere because we saved money here; etc. But there is a more important story behind the multiplication of machines: if the people around me are replaced by machines, it means that i will interact less with humans. Every time a human is replaced by a machine, it decreases the interaction that i will have with other humans. We talk a lot about human-machine interaction, and tend to ignore the fact that a consequence of human-machine interaction is the decline of human-human interaction.

This trend has been going on for at least a century. There used to be armies of telephone operators to direct phone calls, there used to be armies of secretaries typing documents, there used to be armies of sales people serving the customers, etc. Today these human mediators have disappeared and we are increasingly alone in a world of machines.

This trend will continue into the age of Artificial Intelligence to the point that many individuals, especially the older ones, will only interact with machines. Machines will take care of our house, of our errands, of our health, of our entertainment. This will dramatically reduce our need to interact with other human beings; even with our own family, as family support will become less and less necessary.

Your co-workers will be robots. Your friends will be robots. Maybe your lovers will be robots. Your last friend, who will see you die, will be a robot. In many hospitals around the world the last one to take care of a patient on her dying bed is already a machine.

What happens to humanity when you don't interact with humans anymore?

Transcendental Intermezzo: You are a Robot - The Demise of Free Will

I suspect that the program of Artificial Intelligence will have (is having) a powerful effect on the human condition. The shocking revelation coming from A.I. is not that machines can do everything that we can do (we could easily live with this one) but in the specular realization: that we can only do what a machine can do; i.e. that we are just a machine. Robots don't have free will. You can always find the electromechanical cause of a robot's action. You can backtrack and find out the exact sequence of events that triggered the action of a machine. It only depends on how far back in time you want to go, but there is a clear path of causes that have had effects that eventually resulted in that machine lifting its arm or saying, "This life has no meaning". Once these machines start doing everything that we do, it becomes obvious that our actions too are simply the effects of events outside our control.

Think of memory. Humans knew that their memory was fallible, but machines show us on a daily basis how pathetic our memory is. We are literally helpless without a computer memory to remind us of our appointments, our important data, our documents, our favorite videos, etc. We are being diminished by the constant reminder that comes from using a machine to do the things that humans do.

Neuroscience already knew that our actions are caused by electrochemical reactions in the brain, but most of us have always ignored the literal meaning of this finding. Free will is an illusion: even my thoughts about free will are due to neural events in my brain that are beyond my control.

Robots are machines guided by an operating system, by some training, by a program and by some inputs. We are machines guided by genes, by an upbringing, by the ideas of our times, and by the events of our times. Neurological diseases and nutritional deficiencies can alter our behavior just like a software bug or a power outage can affect a robot's behavior.

Evil is, ultimately, due to a combination of factors (genes, upbringing, ideologies, life events) over which we don't have any control just like a robot has no control over how it was built and programmed. Good and evil are meaningless: we are simply machines that are programmed by external forces. A serial killer is no more guilty than an earthquake. The very feelings of morality, revenge, justice and so forth arise from neural processes in the brain that are due to genes, life events and external influences.

Robots will remind us every single second that we are just machines, made of flesh and blood; ultimately, just machines like the machines made of plastic and silicon.

I suspect that, far from leading us to more exciting levels of understanding, the rise of intelligent machines will be a humbling experience for humankind. Copernicus showed us that we are not at the center of the universe. Newton and Einstein showed us that the future is predetermined by the past. Artificial Intelligence will show us that we are not even in control of our actions: we are robots, just like "them".

Transcendental Intermezzo Revised: You are a…?

Science explained too much, even things that we didn't know had to be explained. Science found out that there is a multitude of equations at work that regulate every event in the universe. Science introduced determinism in our thinking: maybe everything is already planned and inevitably proceeds from the initial conditions according to some formulas, and we are just cogs in this giant clockwork.

But we could live with that. Say what you will of me being driven by physical equations, i still feel otherwise.

However, when machines get to the point that they can do everything that we can do… well, it will be harder to feel that there is something special about us. It will be proven that we are just machines.

Underlying the project of Artificial Intelligent is precisely the tacit belief that humans are machines. If we think that we are not machines, we can have a philosophical discussion about the possibility that machines can become human like us. If we think that we are just machines, the discussion is not philosophical: it is technological. It becomes a problem of reverse engineering.

Therefore, when machines finally match all our capabilities, technology will have proven conclusively what Science only hinted at. All those equations failed to convince us that we are mere machines, but technology may prove it beyond any reasonable doubt by showing us a machine that can do everything we can and even better. Then what?

Maybe we do need a new religion to give meaning to this bundle of pulsating organs wrapped in skin that is me.

Why the Singularity is a Waste of Time and Why we Need A.I. – A Call to Action

The first and immediate reason why obsessive discussions about the coming of machine super-intelligence and human immortality are harmful is that they completely miss the point.

We live in an age of declining innovation. Fewer and fewer people have the means or the will to become the next Edison or Einstein. The great success stories in Silicon Valley (Google, Facebook, Apple) are of companies, started by individuals with very limited visions, that introduced small improvements over existing technologies. Entire nations (China and India, to name the obvious ones) are focusing on copying, not inventing.

Scholars from all sorts of disciplines are discussing the stagnation of innovation. A short recent bibliography: Tyler Cowen's e-book "The Great Stagnation" (2010) by an economist; Neal Stephenson's article "Innovation Starvation" (2011) by a sci-fi writer; Peter Thiel's article "The End of the Future" (2011) by a Silicon Valley venture capitalist; Max Marmer's "Reversing The Decline In Big Ideas" (2012) by another Silicon Valley entrepreneur; Jason Pontin's "Why We Can't Solve Big Problems" (2012) by a technology magazine editor; Rick Searle's article "How Science and Technology Slammed into a Wall and What We Should Do About It" (2013) by a political scientist; "The Innovation Illusion" (2016) by economist Fredrik Erixon and investor Björn Weigel. Robert Gordon's book "The Rise and Fall of American Growth" (2016) shows that the pace of innovation has slowed since 1970 after a rapid rise between 1800 and 1970.

Then there is the fundamental issue of priorities. The hypothetical world of the Singularity distracts us from the real world. The irrational exuberance about the coming Singularity distracts a lot of people from realizing the dangers of unsustainable growth, dangers that may actually wipe out all forms of intelligence from this planet.

Let's assume for a second that climate scientists like Paul Ehrlich and Chris Field (i met both in person at Stanford) are right about the coming apocalypse. Their science is ultimately based on the same science that happens to be right about what that bomb would do to Hiroshima (as unlikely as Einstein's formula may look), that is right about what happens when you speak in that rectangular device (as unlikely as it may seem that someone far away will hear your voice), that is right about what happens when someone broadcasts a signal in that frequency range to a box sitting in your living room (as unlikely as it may seem that the box will then display the image of someone located far away), that is right about what happens when you turn on that switch (as unlikely as it is that turning on a switch will light up a room); and it's the same science that got it right on the polio vaccine (as unlikely as it may look that invisible organisms cause diseases) and many other incredible affairs.

The claims about the Singularity, on the other hand, rely on a science (Artificial Intelligence) whose main achievement has been to win board games. One would expect that whoever believes wholeheartedly in the coming of the Singularity would believe tenfold stronger that the human race is in peril.

Let's assume for a second that the same science that has been right on just about everything that it predicted is also right on the consequences of rapid climate change and therefore the situation is exactly the opposite of the optimistic one based mostly on speculation depicted by A.I. science: the human race may actually go extinct before it even produces a single decent artificial intelligence.

In about one century the Earth's mean surface temperature has increased by about 0.8 degrees. Since it is increasing faster today than it was back then, the next 0.8 degrees will come even faster, and there is widespread agreement that 2 degrees above what we have today will be a significant tipping point. Recall that a simple heat wave in summer 2003 led to 15,000 deaths in France alone. Noah Diffenbaugh and Filippo Giorgi (authors of "Heat Stress Intensification in the Mediterranean Climate Change Hotspot ", 2007) have created simulations of what will happen to the Earth with a mean temperature 3.8 degrees above today's temperature: it would be unrecognizable. That temperature, as things stand, is coming for sure, and coming quickly, whereas super-intelligence is just a theoretical hypothesis and, in my humble opinion, is not coming any time soon.

Climate scientists fear that we may be rapidly approaching a "collapse" of civilization as we know it. There are, not one, but several environmental crises. Some are well known: extinction of species (with unpredictable biological consequences: one such is the declining population of bees which may pose a threat to fruit farms), pollution of air and water, epidemics, and, of course, anthropogenic (human-made) climate change. See the "Red List of Threatened Species" published periodically by the International Union for Conservation of Nature (IUCN). See the University of North Carolina's study "Global Premature Mortality Due To Anthropogenic Outdoor Air Pollution and the Contribution of Past Climate Change" (2013) that estimated air pollution causing the deaths of over two million people annually. A Cornell University study led by David Pimentel, "Ecology of Increasing Diseases" (2007), estimated that water, air and soil pollution account for 40% of worldwide deaths. A 2004 study by the Population Resource Center found that 2.2 million children die each year from diarrhea caused by contaminated water and food. And, lest we think that epidemics are a thing of the past, it is worth reminding ourselves that AIDS (according to the World Health Organization) has killed about 35 million people between 1981 and 2012, and in 2012 about 34 million people were infected with HIV (Human Immunodeficiency Virus, the cause of AIDS), which makes it the fourth worst epidemic of all times. Cholera, tuberculosis and malaria are still killing millions every year; and "new" viruses routinely pop up in the most unexpected places (Ebola, West Nile virus, Hantavirus, Avian influenza, Zika virus, etc).

Some environmental crises are less advertised but no less terrifying. For example, global toxification: we filled the planet with toxic substances, and now the odds that some of them interact/combine in some deadly runaway chemical experiment never tried before are increasing exponentially every year. Many scientists point out the various ways in which humans are hurting our ecosystem, but few single out the fact that some of these ways may combine and become something that is more lethal than the sum of its parts. There is a "non-linear" aspect to what we are doing to the planet that makes it impossible to predict the consequences.

The next addition of one billion people to the population of the planet will have a much bigger impact on the planet than the previous one billion. The reason is that human civilizations have already used up all the cheap, rich and ubiquitous resources. Naturally enough, humans started with the cheap, rich and ubiquitous ones, whether forests or oil wells. A huge amount of resources is still left, but those will be much more difficult to harness. For example, oil wells have to be much deeper than they used to. Therefore one liter of gasoline today does not equal one liter of gasoline a century from now: a century from now they will have to do a lot more work to get that liter of gasoline. It is not only that some resources are being depleted, but even the resources that will be left are, by definition, those that are difficult to extract and use (a classic case of "diminishing margin of return").

The United Nations' "World Population Prospects" (2013) estimated that the current population of 7.2 billion will reach 9.6 billion by 2050, and population growth will mainly come from developing countries, particularly in Africa: the world's 49 least developed countries may double in size from around 900 million people in 2013 to 1.8 billion in 2050.

A catastrophic event is not only coming, but the combination of different kinds of environmental problems makes it likely that it is coming even sooner than the pessimists predict and in a fashion that we cannot quite predict.

For the record, the environmentalists are joined by an increasingly diversified chorus of experts in all sorts of disciplines. For example, Jeremy Grantham who is an economist (managing 100 billion dollars of investments). His main point (see, for example, his 2013 interview on Charlie Rose's television program) is that the "accelerated progress" that the Singularity crowd likes to emphasize started 250 years ago with the exploitation of coal and then truly accelerated with the exploitation of oil. The availability of cheap and plentiful energy made it possible to defy, in a sense, the laws of Physics. Without fossil fuels the human race would not have experienced such dramatic progress in merely 250 years. Now the planet is rapidly reaching a point of saturation: there aren't enough resources for all these people. Keeping what we have now is a major project in itself, and those who hail the coming super-intelligence miss the point the way a worker planning to buy a bigger house misses the point that he's about to get fired.

We are rapidly running out of cheap resources, which means that the age of steadily falling natural resource costs is coming to an end. In fact, the price of natural resources declined for a century until about 2002 and then in just 5 or 6 years that price regained everything that it had lost in the previous century (i am still quoting Grantham). This means that we may return to the world of 250 years ago, before the advent of the coal (and later oil) economy, when political and economic collapses were the norm; a return to, literally, the ages of starvation.

It is not only oil that is a finite resource: phosphates are a finite resource too, and the world's agriculture depends on them.

Population growth is actually a misleading parameter, because "overpopulation" is measured more in terms of material resources than in number of people: most developed countries are not overcrowded, not even crowded Singapore, because they are rich enough to provide a good life to their population; most underdeveloped countries are overcrowded because they can't sustain their population. In this sense, overpopulation will increase even in countries where population growth is declining: one billion Indians who ride bicycles is not the same as one billion Indians who drive cars, run A/C units and wrap everything in plastic. If you do it, why shouldn't they?

The very technologies that should improve people's lives (including your smartphone and the robots of the future) are likely to demand more energy, which for now comes mainly from the very fossil fuels that are leading us towards a catastrophe.

All those digital devices will require more "rare earths", more coltan, more lithium and many other materials that are becoming scarcer.

We also live in the age of Fukushima, when the largest economies are planning to get rid of nuclear power, which is the only form of clean alternative energy as effective as fossil fuels. Does anyone really think that we can power all those coming millions of robots with wind turbines and solar panels?

Chris Field has a nice diagram (expanded in the 2012 special report of the Intergovernmental Panel on Climate Change titled "Managing the Risks of Extreme Events and Disasters to Advance Climate Change Adaptation") that shows "Disaster Risk" as a function of "Climate Change" and "Vulnerability" (shown, for example, at a seminar at the Energy Biosciences Institute in 2013). It is worth pondering the effects of robots, A.I. and the likes on that equation. Manufacturing millions of machines will have an impact on anthropogenic climate change; economic development comes at the cost of exploitation of finite resources; and, if high technology truly succeeds in increasing the longevity of the human race, the population will keep expanding. In conclusion, the race to create intelligent machines might exacerbate the risk of disasters before these super-intelligent machines can find a way to reduce it.

The Paris accord on climate change of 2015 (COP21) was a wildly optimistic agreement, and not even an enforceable one.

Economists such as Robin Hanson ("Economics Of The Singularity", 2008) have studied the effects of the agricultural, industrial and digital revolutions. Each caused an acceleration in economic productivity. The world's GDP may double every 15 years on average in this century. That's an impressive feat, but it's nothing compared with what would happen if machines could replace people in every single task. Productivity could then double even before we can measure it. The problem with that scenario is that the resources of the Earth are finite, and most wars have been caused by scarcity of resources. Natural resources are already strained by today's economic growth. Imagine if that growth increased ten fold, and, worse, if those machines were able to mine ten or 100 times faster than human miners. It could literally lead to the end of the Earth as a livable planet. Imagine a world full of machines that rapidly multiply and improve, and basically use all of the Earth's resources within a few years.

Ehrlich calls it "growthmania": the belief that there can be exponential growth on a finite planet.

The optimists counter that digital technology can be "cleaner" than the old technology. For example, the advent of email has dramatically reduced the amount of paper that is consumed, which has reduced the number of trees that we need to fell. It is also reducing the amount of mail trucks that drive around cities to deliver letters and postcards. Unfortunately, in order to check email and text messages you need devices like laptops, notepads and smartphones. The demand for materials such as lithium and coltan has risen exponentially.

Technological progress in the internal combustion engine (i.e., in fuel-efficient vehicles), in hybrid cars, in electric cars and in public transportation is credited for the reduction in oil consumption since 2007 in developing countries. But Asia Pacific as a whole has posted a 46% increase in oil consumption in the first decade of the 21st century. In 2000 oil consumption in China was 4.8 million bpd (barrels per day), or 1.4 barrels per person per year. In 2010 China's consumption had grown to 9.1 million bpd. China and India together have about 37% of the world's population. The rate of cars per person in China (0.09%) is almost 1/10th the one in the USA (0.8%) and in India is one of the lowest in the world (0.02%). Hence analysts such as Ken Koyama, chief economist at the Institute of Energy Economics Japan, predict that global petroleum demand will grow 15% over the next two decades ("Growing Oil Demand and SPR Development in Asia", 2013).

George Mitchell pioneered fracking in 1998, releasing huge amounts of natural gas that were previously thought inaccessible. Natural gas may soon replace oil in power stations, petrochemical factories, domestic heaters and perhaps motor vehicles. The fact that there might be plenty of this resource in the near future proves that technology can extend the life expectancy of natural resources, but it does not change the fact that those resources are finite, and it might reduce the motivation to face the inevitable.

Technology is also creating a whole new biological ecosystem around us, a huge laboratory experiment never tried before. Humans have already experienced annihilation of populations by viruses. Interestingly, the three most famous ones took hold at a time of intense global trade: the plague of 1348 (the "black death") was probably brought to Europe by Italian traders who picked it up in Mongol-controlled regions at a time when travel between Europe and Asia was relatively common and safe; and the flu pandemic of 1918, that infected about 30% of the world's population and killed 50 million people, took hold thanks to the globalized world of the British and French empires and to World War I. The HIV came out in the 1980s when the Western economies had become so entangled and it spread to the whole world during the globalization decade of the 1990s. By the end of 2012 AIDS had killed 35 million people worldwide.

We now live in the fourth experiment of that kind: the most globalized world of all times, in which many people travel to many places; and they do so very quickly. There is one kind of virus that could be worse than the previous ones: a coronavirus, whose genes are written in RNA instead of DNA. The most famous epidemics caused by a coronavirus was the Severe Acute Respiratory Syndrome (SARS): in February 2003 it traveled in the body of a passenger from Hong Kong to Toronto, and within a few weeks it had spread all over East Asia. Luckily both Canada and China were equipped to deal with it and all the governments involved did the right thing; but we may not be as lucky next time. In 2012 a new coronavirus appeared in Saudi Arabia, the Middle East Respiratory Syndrome (MERS).

All of these race-threatening problems are unsolved because we don't have good models for them. One would hope that the high-tech industry invest as much into creating good computational models that can be used to save the human race as into creating ever more lucrative machines. Otherwise, way before the technological singularity happens, we may enter an "ecological singularity".

Discussing super-human intelligence is a way to avoid discussing the environmental collapse that might lead to the disappearance of human intelligence. We may finally find the consensus to act on environmental problems only when the catastrophe starts happening. Meanwhile, the high-tech world will keep manufacturing, marketing and spreading the very items that make the problem worse (more vehicles, more electronic gadgets, and, soon, more robots); and my friends in Silicon Valley, firmly believing that we are living in an era of accelerating progress, will keep boasting about the latest gadgets... the things that environmental scientists call "unnecessarily environmentally damaging technologies".

Fans of high technology fill their blogs with news of ever more ingenious devices to help doctors, not realizing that the proliferation of such devices will require even more energy and cause even more pollution (of one sort or another). They might be planning a world in which we will have fantastic health care tools but we will all be dead.

I haven't seen a single roadmap that shows how technology will evolve in the next decades, leading up to the Singularity (to super-human intelligence). I have, instead, seen many roadmaps that show in detail what will happen to our planet under current trends.

There is also plenty to worry about the Internet. As Ted Koppel wonderfully explained in his book "Lights Out" (2015), the chances of a massive cyber-attack, that would leave the USA without electricity, communications and even water for weeks, are very high. There are dozens of hacking incidents every day. Banks, retail chains, government agencies, even the smartphone of the director of the CIA and even Mark Zuckerberg's Facebook account have been hacked. And they are usually hacked by amateurs in search of publicity. Spy agencies can cause a lot more damage than amateurs. They are probably monitoring the system right now, and they will strike only when it is worth their while. Companies that boasted about being invulnerable to hacking attacks have frequently been subjected to humiliating hacking attacks. The fact is that the Internet cannot be defended. It was probably a strategic mistake to make so much of the economy and of the infrastructure depend on a computer network (any computer network). Computers are vulnerable in a way that humans are not. You need to capture me and torture me in order to extract information from me that would harm my friends, relatives and fellow citizens; but you don't need to capture and torture a computer. It is much easier than that. Computer networks can be easily fooled into providing access and information. The more intelligent you make the network of computers, the bigger the damage it can cause to the humans who use it.

A.I.'s promises of dramatic economic and social change have been very effective in obtaining public and private funding, but that has come at the expense of other disciplines. Steven Weinberg's book "Dreams of a Final Theory" (1993) failed miserably to convince the political establishment to fund a new expensive project, the Superconducting Super-Collider. He failed because he narrated the reality of scientific research. Ray Kurzweil's "The Age of Spiritual Machines" (1999), a provocative and enthusiastic (and wildly self-congratulatory) reaction to IBM's Deep Blue beating the world chess champion in 1997, was totally out of touch with reality but impressed the political establishment enough such that many A.I. scientists obtained funding for their research. Research in A.I. in the USA has always relied on funding from the government (mainly through its "defense" arm called DARPA, which is really a designer of weapons). It was true of the original A.I. labs at the MIT and Stanford, it was true of the A.I. research at SRI that yielded the autonomous robot Shakey and eventually the conversational agent Siri, and it was true of Nicholas Negroponte's Media Lab at the MIT. Capturing the imagination of the political and military establishment is imperative for the progress of a scientific program (in Europe a similar phenomenon is at work, although it is the social impact rather than the military one to be more valued). The media's passion for A.I. may end up draining legitimate disciplines of the funding they need to improve the lives of millions of people. Imagine of enthusiasm for early A.I. had diverted the funds that were spent on the Interstate Highway System or Social Security; or if today's enthusiasm ends up diverting some of the $7 billion that the government pays to the Center for Disease Control (CDC), our front line in fighting infectious diseases. I for one think that, in the grand scheme of things, the Superconducting Super-Collider would have been more useful than Siri.

Last but not least, we seem to have forgotten that a nuclear war (even if contained between two minor powers) would shorten the life expectancy of everybody on the planet, and possibly even make the planet uninhabitable. Last time I checked, the number of nuclear powers had increased, not decreased, and, thanks to rapid technological progress and to the electronic spread of knowledge, there are now many more entities capable of producing nuclear weapons.

The unbridled optimism of the Artificial Intelligence community, and of the media that propagate it, is not justified because A.I. is not helping to solve any of these impelling problems. We desperately need machines that will help us solve these problems. Unbridled optimism is not a replacement for practical solutions.

The enthusiastic faith that Rome was the "eternal city" and the firm belief that Venice was the "most serene republic" did not keep those empires from collapsing. Unbridled optimism can be the most lethal weapon of mass destruction.

The Future of Human Creativity

Maybe we should focus on what can make us (current Homo Sapiens people) more intelligent, instead of focusing on how to build more intelligent machines that will make our intelligence obsolete. Creativity is what truly sets Homo Sapiens apart from other species.

There are two myths here that i never bought. The first one is that adults are more intelligent than children, and therefore children have to learn from adults, not vice versa.

Children perform an impressive feat in just a few years, acquiring an incredible amount of knowledge and learning an incredible portfolio of skills. They are also fantastically creative in the way they deal with objects and people. Teenagers are still capable of quick learning (for example, foreign languages) and can be very creative (often upsetting parents and society that expect a more orthodox behavior, i.e. compliance with rules). Adults, on the other hand, tend to live routine lives and follow whatever rules they are told to obey.

When i look at the evidence, it seems to be that creativity, and therefore what is unique about human intelligence, declines with age. We get dumber and less creative, not smarter and more creative; and, once we become dumb adults, we do our best to make sure that children too become as dumb as us.

Secondly, the people of the rich developed high-tech world implicitly assume that they are more intelligent and creative than the people of the poor undeveloped low-tech world. In my opinion, nothing could be farther from the truth. The top of creativity is encountered in the slums and villages of the world. It is in the very poor neighborhoods that humans have to use their brain every single minute of their lives to come up with creative and unorthodox solutions, solutions, that nobody taught them, to problems that nobody studied before. People manage to run businesses in places where there is no infrastructure, where at any time something unpredictable can (and will) happen. They manage to sell food without a store. They manage to trade without transportation. When they obtain a tool, they often use it not for the purpose for which it was originally designed but for some other purpose. They devise ever new ways to steal water, electricity, cable television and cellular phone service from public and private networks. They find ways to multiply and overlap the functions of the infrastructure (for example, a railway track also doubles as a farmer's market, and a police road-block becomes a snack stop). They help each other with informal safety networks that rival state bureaucracies (not in size or budget, but in effectiveness). The slums are veritable laboratories where almost every single individual (of a population of millions) is a living experiment (in finding new ways of surviving and prospering). There is no mercy for those who fail to "create" a new life for themselves every day: they stand no chance of "surviving".

If one could "measure" creativity, i think the slums of the world would easily outperform Silicon Valley.

Robots will replace Silicon Valley engineers way before they can replace the humble seller of pillows at the bus station who walks around barefoot trying to locate the most likely customer among the thousands of frantic long-distance passengers.

These highly creative people yearn for jobs in the "white" economy, the economy of the elite that lives outside the slums. For that "white" economy they may perform trivial repetitive jobs (chauffeur, cashier, window washer); which means that they have to leave their creativity at home. The "white" economy has organized daily life in such a way (around "routines") that everybody is guaranteed to at least survive. The people of the slums use their brains only when they live and work in the slums. When they live or work outside the slums, they are required to stop being creative and merely follow procedures, procedures that were devised by vastly less creative people who would probably not survive one day in the slums. Routines maximize productivity precisely by reducing human creativity. Someone else "creates", and the worker only has to perform, a series of predefined steps. The slum dweller cannot be replaced by a machine, but the "routinized" worker can be.

The routine, however, is useful for businesses because it can "amplify" the effect of innovation. The innovation may be very small and very infrequent, but the effect of the routine performed by many workers (e.g., by many Silicon Valley engineers) is to make even the simplest innovation relevant for millions of individuals.

The creativity of slums and villages, on the other hand, is constant, but, lacking the infrastructure to turn it into routine, ends up solving only a small problem for a few individuals. The slums are a colossal reservoir of creative energies that the world is wasting, and, in fact, suppressing.

In our age we are speeding up the process by which (rule-breaking) children become (rule-obeying) adults and, at the same time, we are striving to turn the creativity of the slums into the routine of factories and offices. It seems to me that these two processes are more likely to lead to a state of lower rather than higher intelligence for the human race.

I suspect that removing the unpredictable from life means removing the very essence of the human experience and the very enabler of human intelligence. On the other hand, removing the unpredictable from life is the enabler of machine intelligence.

Why I am not Afraid of A.I.

Between 2014 and 2015 Silicon Valley's serial entrepreneur Elon Musk and British physicist Stephen Hawking, as well as the richest man in the world, Bill Gates, all sounded alarm bells about the danger posed to humankind by Artificial Intelligence. They were descendants of Bill Joy's "The Future doesn't need us". In 2016 Elon Musk and Peter Thiel founded OpenAI, a non-profit organization with the mission "to advance digital intelligence in the way that is most likely to benefit humanity as a whole". They hired Ilya Sutskever, formerly at Google and in Hinton's group, to lead the research, and hired advisors such as Pieter Abbeel of UC Berkeley, Yoshua Bengio, and personal-computer pioneer Alan Kay.

I, instead, am not afraid of A.I. because we are not even remotely close to having truly intelligent machines.

I am not afraid of A.I.: I am afraid that it will not arrive soon enough. Machines are essential to our well-being today, and will increasingly determine our well-being in the future, and more intelligent machines are probably indispensable to solve many of the gravest problems of our era.

A world without robots is a world in which humans have to work for very low wages in order to produce goods that ordinary families can afford to buy. It is a world in which only the rich could afford a car or even a TV set. A world without robots is a world in which humans would have to perform all sorts of dangerous and unhealthy jobs, such as cleaning up Fukushima's nuclear disaster, and would have to work in horrible conditions inside mines and steelworks. Robots are being used to disarm suicide bombers and to remove landmines. Without robots, these tasks would be carried out by human beings. A world without robots would be a terrible world.

Robots suffer from poor marketing. Robots are mostly presented as big, scary beasts. We should instead publicize the fact that someday the hardware store next door will offer tiny robots capable of crawling inside the plumbing of our house and of unclogging the pipes. Robotic "exosuits" will allow us to lift and carry heavy weights in the backyard. And so forth: robots will help us by solving practical problems around the house.

Do we need service robots? Do they steal jobs? Have you ever desperately looked for a human being in a large store, a shopping mall, a hotel lobby, a train station, a public office, or even in a street? How many times does it happen that you have a simple question and there's nobody to talk to? Or maybe there are people but they don't speak your language, or they are tourists like you, or they are wearing headsets and listening to music? Wouldn't it be nice if a robot could answer your question and, if it is about a location, even take you there? It could even be that humankind finally gets rid of the "hours of operation". Service robots can someday keep a store open nonstop even when all the human staff is asleep. That would be a much more dramatic achievement than super-intelligent machines.

These service robots do not steal jobs. Those jobs today don't exist. Maybe they existed a century ago, and maybe they still exist in poor countries. I have been in developing countries where a clerk welcomes you when you enter the store and gladly helps you to find the product that you are looking for. But this has become a rarity in developed countries.

Instead of worrying about the jobs that may be "stolen" by machines, we should worry about the jobs that we will soon need and for which we are not prepared. Taking care of elderly people is a prime example. There is virtually no country where population growth is accelerating. In most countries of the world, population growth is decelerating. In some countries it is turning negative. In many countries the population has peaked and will soon start decreasing while at the same time aging thanks to improvements in medicine. In other words, many countries need to prepare for a future with a lot of older people with fewer younger people who can take care of them. In the Western world the 1950s and 1960s were the age of the "baby boom". The big social revolution of the 21st century will be the boom of elderly people. The rich world is entering the age of the "elderly boomers". Who is going to take care of that aging population? Most of these aging people will not be able to afford full-time human care. The solution is robots: robots that can go shopping, that can clean the house, that can remind people to take their medicines that can check their blood pressure, etc. Robots could do all of these things day and night, every day of the year, and at an affordable price. I am afraid that A.I. will not come soon enough and we will face the aging apocalypse.

We profess that we want all the people in the world to become rich like the rich Western countries, but the truth is that any "rich" society needs poor people to perform all the vital jobs that the "richer" people refuse. Poor people take care of most of the chores that keep society working and that keep us alive. Those are humble and low-paid jobs such as collecting garbage and making sandwiches. We profess that we want all eight billion people of the planet to have the same standard of living that the rich world has, but what happens when all eight billion people become rich enough that nobody wants to take those humble and low-paid jobs? Who is going to collect the garbage once a week, who is going to make sandwiches at the lunch cafeteria, who is going to clean the public bathrooms, who is going to wash the windows of the office buildings? We don't want to admit it, but today we rely on the existence of millions of poor people who are willing to do those jobs that we don't want to do. I hope that we will indeed solve the problem of poverty in 50 years or even less; what that means is that we only have 50 years to invent robots that can do all the jobs that people will not want to do 50 years from now. I am not scared of robots, i am scared of what will happen in 50 years if we don't have intelligent robots to collect garbage, make sandwiches, clean bathrooms, etc.

A world without robots is a dysfunctional world, a world of very poor people working and living in horrible conditions, a world of societies that cannot care for the elderly and that cannot help people with permanent disabilities.

A world without robots is a scary place.

A Conclusion: Religion in the Age of A.I.

Humans have been expecting a supernatural event of some kind or another since prehistory. Human brains seem to be programmed to believe in the supernatural and to strive for immortality.

As mentioned at the beginning of this book, we are witnessing the birth of a new religion, a religion that believes in a supernatural world that exists not in this universe, not in the heavens, but in the dataverse.

According to this new "religion", A.I. will generate a kind of supernatural intelligence that will rule over the human world.

In retrospect, ancient religions were realistic: they admitted that we all have to die and looked for hope in the afterlife. This was an empirical and rational approach. The narrative of the Singularity denies the obvious: that everything has an ending. This is neither empirical (there is no evidence of eternal immortal beings) nor rational (there is no science that would justify something living longer than the lifetime of the universe, or, for that matter, just the lifetime of the Sun).

Rationality is under attack from both the right and the left, the "right" being modern spirituality and the "left" being the Singularity camp.

The modern spirituality (the "right") largely rejects the superstitions of Judaism, Christianity and Islam in favor of Daoism and Buddhism, the two philosophies (not quite religions) that seem to best match what we know about the universe; and the two philosophies that have been prominent in the San Francisco Bay Area since the 1960s, i.e. the two that Silicon Valley naturally encountered as it abstracted technology into a philosophy of life. The "right" was born approximately with Fritjof Capra's "The Tao of Physics" (1975), written by a physicist, and Michael Singer's bestseller "The Untethered Soul" (2007) or William Broad's "The Science of Yoga" (2012) are typical examples of its mature stage, leaving behind attempts to merge spirituality and physics such as Deepak Chopra's "Quantum Healing" (1989) and Danah Zohar's "The Quantum Self" (1990), co-written with a physicist.

The "left" is part of a more general movement that thinks of the universe as ruled by data. Israeli historian Yuval Harari calls it a new religion, Dataism, in his book "Sapiens - A Brief History of Humankind" (2011). This "left" has faith in science, and notably in computer science. According to the Pew Research Center, in 2015 a whopping 89% of US adults believed in some kind of god (in Europe we estimate that number to be about 77%), but very few of them are looking forward to the afterlife. Most of them are terrorized by the idea of dying. Their real religion is medicine, and, indirectly, science. They tolerate scientific research in chemistry, biology, physics and the like because they hope that it will contribute to progress in medicine. They were happy that in 1993 the government killed an expensive project to build the world's fastest particle accelerator (the "Superconducting Supercollider") and that in 2011 the government shut down the most powerful particle accelerator in the country (the Tevatron); but tell them that accelerating hadrons will prolong their lives and they will gladly pay taxes for the most expensive particle accelerator ever built. Tell them that space exploration will prolong their lives and they will gladly pay taxes for a mission to Saturn. Tell them that Artificial Intelligence will grant them immortality, and that a super-human intelligence (the Singularity) is truly coming soon, and they will react the way Jews, Christians and Muslims used to react to the news that the Messiah is coming: fear and hope; fear that the Messiah will send us to hell (a well-deserved hell, judging from my favorite history books) and hope that the Messiah will be so merciful as to grant us immortality.

I also submit that much of the "exponential progress" that we witness today is due to the retreat of religious institutions. Religious institutions, whether in the Catholic world or in the Islamic world, mostly resisted governments. They would not support and sometimes not even protect scientists, engineers, philosophers and physicians who hinted in any way that the soul does not exist or that it is the mere manifestation of electrochemical brain processes. Religion is naturally hostile to technological and scientific progress because progress distracts from the foundation of religious morality, the soul, besides clearly upsetting the traditional social life upon which priests rely for their power. Therefore, we live inside a positive-feedback loop: religions decline, their decline fosters scientific and technological progress, progress causes religion to decline, etc. No wonder that belief in spiritual superbeings is rapidly being replaced by belief in a technological superbeing.

Traditional religions worked well when there was no hope for a remedy to death. Hence the only remedy was faith in a god's mercy. The new technological religion offers a terrestrial remedy to death: terrestrial longevity if not immortality. Whether this new religion is any more realistic than the ancient Western religions is debatable; and whether the Singularity or the Messiah or nobody comes to our rescue, remains to be seen.

The Singularity may become the new religion for the largely atheistic crowd of the high-tech world. Just like with Christianity and Islam, the eschatological mission then becomes how to save oneself from damnation when the Singularity comes, balanced by the faith in some kind of resurrection.

We've seen this movie before, haven't we?

P.S. What i did not Say

I want to emphasize what i did not say in this book.

I did not claim that an artificial general intelligence is impossible, (only that it requires a major revolution in the field); and i certainly did not argue that superhuman intelligence is not possible (in fact, i explained that it is already all around us); and i did not rail against technological progress (i lamented that its achievements and benefits are wildly exaggerated).

I did not write that technology makes you stupid. I wrote that rules and regulations make you stupid; and technology is used to establish, enforce and multiply those rules and regulations (often giving machines an unfair advantage over humans).

I did not write that there has been no progress in Artificial Intelligence: there has been a lot of progress, but mainly because of cheaper, smaller and faster processors, and because of better structured environments in which it is easier to operate (both for humans and for machines).

I did not write that humans will never create an artificial intelligence. We have already created Artificial Intelligence programs that do many useful things, as well as some truly obnoxious ones (like displaying ads on everything you do online). The definition of "intelligence" is so vague that the very first computer (or the very first clock) can be considered an artificial intelligence. In fact, early computers were called "electronic brains", not "electronic objects".

I did not write that Artificial Intelligence is useless. On the contrary, i think that it has helped neuroscience. Maybe the "enumeration" problem (the problem of enumerating all the intelligent tasks that are needed in order to achieve artificial general intelligence) is a clue that our own brain might not be "one" but a confederation of many brains, each specialized in one intelligent task.

I did not write that i am afraid of intelligent machines. Quite the opposite: we desperately need intelligent machines. Technological progress has had many downsides but, overall, it has helped humans live better lives. We do need the progress in machine intelligence that technology has promised and not delivered.

I am not afraid of intelligent machines. I am afraid that it will not come soon enough.

Appendix: The Dangers of Clouding - Wikipedia as a Force for Evil

Since 2001, when Jimmy Wales started it, Wikipedia has grown to become the largest cooperative project in the history of the world, with 39 million articles in 250 languages (as of 2013). The jury, however, is still out on whether Wikipedia is a force for good or a force for evil.

The traditional debate over Wikipedia has focused on how much we can trust thousands of anonymous editors as opposed to the small team of highly decorated scholars who curate the traditional encyclopedia. Since scholars and erudite people in general are less likely to get into a brawl, the fear was that in the long run the "mob" would win, the frequent outcome in many facets of popular culture. That was pretty much the only concern when Wikipedia was just that: a substitute for the encyclopedia.

However, the Internet is not a bookshelf. Those who treat the Internet like a bookshelf miss the point about its impact, which is not just to replace existing objects and services.

In mid 2010 i searched Wikipedia for biographies of the main politicians and consistently found adulatory comments, even for those who have been responsible for blatant violations of human rights. In my research for my book on Silicon Valley i flipped through thousands of Wikipedia pages about companies and individuals: the vast majority were simply the equivalent of press releases worded according to the current business strategy of the company or according to the whims of the individual. In late 2010 the article on Feminism presented Mohammed (the founder of Islam) as the first major feminist in the history of the world. In February 2011 the article on detective fiction mentioned the medieval Arabian collection of stories "One Thousand and One Nights" as the first suspenseful book. Wikipedia pages on albums and films routinely describe them with a "Response from the critics was generally positive" comment, completely omitting the devastating reviews published by reliable critics. Obviously, these were all cases in which someone with a specific agenda was trying to influence the millions of people who rely on Wikipedia.

I started noticing a disturbing fact: the popularity of Wikipedia is de facto obliterating all the alternative sources that one could use to doublecheck Wikipedia articles. A search on any major topic routinely returns a Wikipedia page in the first two or three lines. The other lines in the first page of results are almost inevitably commercial in nature. In order to find a scholarly page that can prove or disprove the Wikipedia page, one has to flip through several pages of results. Very few people make the effort. Therefore Wikipedia is rapidly becoming the only source of information about any major topic. Maybe this is acceptable for scientific topics (although i would still prefer that my Quantum Physics and Cellular Biology came from someone who has signed the article with her/his name and affiliation) but it is dangerous for topics that are "politicized" in nature. Then Wikipedia becomes the only source that millions of people access to find out what a politician, a government or a company has done. Worse: every topic can be "politicized" to some extent. I found references to the Bible and the Quran in articles about scientific topics. No traditional encyclopedia and no academic textbook in the free world would reference the Bible or the Quran to explain Quantum Mechanics or Cellular Biology. Precisely because it is edited by the general public, Wikipedia lends itself to a global politicization of every topic. It is an illusion that Wikipedians carry out "anonymous and collaborative editing": the very nature of Wikipedia encourages people to avoid collaboration and rather invites them to suffuse ideological agendas into encyclopedia pages. The "collaboration" of which Wikipedia boasts is the fact that someone can retaliate to an opinionated or biased statement by removing or altering that statement and maybe inserting one that leans in the opposite direction; but a brawl is a very loose definition of "collaboration".

That danger is very visible in the rapid decline of quality. Like any corporation that has to hide its own shortcomings, Wikipedia boasts study after study that shows it to be as accurate and more complete than the Encyclopedia Britannica. This is true only if one ignores semantics. In reality, there has never been and never will be a

Britannica article that is simply the press release from a company or a doctored biography from a tyrannical government. If one considers the semantics, the gap between the accuracy of the traditional encyclopedia and the inaccuracy of Wikipedia is rapidly increasing.

The evil is, obviously, not coming from the founder or the staff. It originates from the success itself of Wikipedia. According to a diagram from a 2011 presentation by Zack Exley that i attended, the number of senior (unpaid) Wikipedia editors rapidly reached 60,000 and has declined a bit during the Great Recession. That number, of course, does not tell the whole story. The meaningful number is the number of pages that an average unpaid editor has to maintain. In 2003 (just before the Wikipedia explosion) there were less than 200,000 articles and about 60,000 editors: on average three pages per senior editor. In 2010 the number of editors declined to 50,000 while the number of articles in English alone had increased to ten million (according to a diagram that is currently posted on the Wikipedia website (http://en.wikipedia.org/wiki/File:EnwikipediaArt.PNG): even assuming that all those 50,000 editors stick to Wikipedia's original philosophy (i'll say later why i don't believe it), it would mean 200 articles on average per editor.

Here is the bigger problem. When there were only a few thousand users, there was little interest from governments and corporations in what Wikipedia said. Now that there are millions of users and the Wikipedia page is usually among the first few presented by a search engine, the interest in determining what Wikipedia displays has grown enormously. There has been an undocumented explosion in the number of Wikipedia editors who are "paid" (either salaried or contracted) by governments, organizations, corporations and celebrities to twist the text of a Wikipedia article so that it represents the interest of that government, organization, corporation or celebrity.

When there were only a few thousand articles, it was relatively easy for the unpaid idealistic editors to control the content of

Wikipedia. Now that there are millions of articles, it is simply impossible for those unpaid idealistic editors to control what the paid editors do. To make matters worse, Wikipedia covets the idea that editors have to be anonymous: therefore there is no way for an unpaid idealistic editor to know if another editor is unpaid or paid. It's like those horror movies in which there is no way for a human to know whether she is surrounded by humans or zombies.

Like any corporation that has to hide its own shortcomings, Wikipedia boasts that "In the month of July 2006, Wikipedia grew by over 30,000,000 words". But that's precisely the problem. That's precisely what is frightening. Many of those 30 million words may be written by unprofessional, biased and sometimes paid "editors" whose interest in creating an encyclopedia is much lower than their interest in promoting a viewpoint or serving their employer. This leaves less than 50,000 unpaid idealistic Wikipedia editors to fight against an increasing number of editors paid by government agencies, ideological organizations, corporations and celebrities, not to mention the thousands of occasional uninformed amateurs who want to shout their opinion to the world.

Needless to say, a government agency, an ideological organization, a corporation or a celebrity has more resources at its disposal, and is much more determined, than a hapless unpaid Wikipedian. Therefore their version of the facts will eventually win. No wonder that an increasing number of pages simply displays what the subject of the page wants people to read. It is pointless for an idealistic editor to fight against it: the corporation or organization interested in that page has overwhelming resources to win the fight. There is no "brawl" over the content of those pages because it would be pointless. The net result is that Wikipedia is inevitably being hijacked by entities whose goal is not to spread knowledge but to spread propaganda.

Furthermore, several governments around the world block Wikipedia webpages. In the Middle East we were not able to access pages about Israel and Islam. In mainland China we could not access just about any page about history, including my own website www.scaruffi.com However, the free world can view the pages that have been doctored by the Chinese government and by Islamic religious groups. Therefore there is a one-way flow of mental conditioning: "their" people cannot see our version of the facts, but we are increasingly exposed to "their" version of the facts as "they" become more and more active in editing Wikipedia pages. It is not difficult to predict who will win in the long run.

For government agencies, ideological organizations, corporations and celebrities Wikipedia has become a fantastic device to brainwash not only your own audience but all the people in the world.

Politically speaking, Wikipedia is de facto a force opposed to the change that social media foster. While Facebook and Twitter cannot be easily hijacked by authorities and corporations to brainwash people with distorted facts, Wikipedia can be and is being used precisely for that purpose by an increasing number of skilled and sinister "editors". Wikipedia can potentially become a force to stop change and promote repression, corruption, speculation and possibly genocide. Because they are so distributed and cannot be "edited", the voices expressed by Facebook and Twitter represent the voice of the people. The centralized Wikipedia, instead, is increasingly representing the voice of the oppressor; or, if you prefer, the oppressors are increasingly keen on appropriating Wikipedia.

In parallel, Wikipedia is having a detrimental effect on culture: it is sending out of business the only sources that we can use to verify Wikipedia's accuracy: the traditional encyclopedias. Compiling an encyclopedia is a colossal endeavor that requires the collective work of dozens of distinguished scholars. The cost for the publisher is enormous. In the age of Wikipedia no publisher is crazy enough to invest millions for an encyclopedia that will have to compete against the much bigger and absolutely free of charge Wikipedia. The age of encyclopedias that began with the Enlightenment in the 18th century is coming to an end in the 21st century. In other words, the fact that Wikipedia is free has created a problem of historical proportions. Since no more encyclopedias will be produced, and any specialized website will be infinitely difficult to find using a search engine, society will have no way to determine if a Wikipedia article is telling the truth or not. There will be no second source where one can double check a statement, a date, a story, let alone discuss the merits of who is represented on Wikipedia and who is not. Wikipedia is sending out of business the very sources that we use to determine Wikipedia's reliability and accuracy, the very sources that we used for centuries to determine the veracity of any statement. Wikipedia is not an encyclopedia, it is becoming a colossal accumulation of propaganda and gossip. The destruction of the traditional encyclopedia may send us back to the dark ages that followed the collapse of the Roman Empire.

P.S.

Things may be getting even more sinister as i write this book. Wikipedia's claim that anybody can edit an article is rapidly becoming an illusion: in reality, millions of IP addresses are banned from editing Wikipedia. A Stanford friend who added a link to a Wikipedia article (linking to this very article of mine) has never been able to edit articles again: Wikipedia displays an error message in which he is accused of "non constructive behavior". If this reminds you of totalitarian regimes, welcome to the world of Wikipedia. Wikipedia, by its own admission, keeps a detailed record of what every IP address in the world has written on which articles. And de facto Wikipedia bans from editing its pages those places (like libraries) that don't allow it to track down the identity of the person by the IP address. This is exactly what secret police like the KGB have always done on behalf of totalitarian regimes in which you are supposed to read (what they want you to read) but not to write (what you would like the world to read).

The usual objection to this comparison of mine is that Wikipedia editors are volunteers who do it just because they believe in the ideal. You'd be surprised how many members of the secret police in places like Nazi Germany, the Soviet Union and today's Iran were and are fanatical volunteers who believed in the ideal of their totalitarian state and were willing to work for free to fight the enemies of that state. The real enemy is often not the dictator in charge but the fanatics who legitimize that dictator. Without those fanatical followers the totalitarian state would collapse.

Most users of Wikipedia have trouble accepting that Wikipedia is bad for humankind. They admit the limits and the potential harm, but would not want to erase it from the Web. My friend Federico Pistono, author of "Robots Will Steal Your Job, But That's OK" (2012), told me: "We just need to educate people on how to use it". My counter-suggestion: we should introduce more mistakes. It is important that the users of Wikipedia get "educated" about the fact that Wikipedia articles are typically biased articles written by whoever has more time and more money to continue editing them. In the interest of the truth, please change an article on the Nazi massacre of Jews in Poland so that "Warsaw" becomes "Acapulco" and "Hitler" becomes "Mickey Mouse". This way people will be aware that they cannot trust an anonymous Wikipedia article and they have to use other sources to doublecheck the content of Wikipedia articles. Sure: Wikipedia is useful to find out that Paris is the capital of France, and that the population of Nigeria is 173 million. It is very "useful" for many purposes. As long as we don't make excessive claims about its reliability: it is NOT an encyclopedia. At best, it is just a collection of the advice given by amateurs to amateurs, just like reviews on Yelp and Amazon. Many television shows, documentaries and Internet videos have been useful in raising awareness about world events, but (hopefully) people know that those shows are run by comedians, entertainers and amateurs. Because Wikipedia articles are anonymous, people are routinely misled into thinking that they were written by top authorities more reliable than comedians and entertainers. In many cases that is not true. In fact, i don't know a single scholar who has contributed even a single article to Wikipedia.

How about a big banner on every Wikipedia article that warns "Disclaimer: None of the texts published here was provided or verified by a competent scholar"? Just like we warn people that cigarettes cause cancer?

Appendix: The Myth of Longevity

The new cult of digital immortality goes hand in hand with the widely publicized increases in life expectancy.

For centuries life expectancy (at older ages) rose very little and very slowly. What truly changed was infant mortality, that used to be very high. But towards the end of the 20th century life expectancy posted an impressive increase: according to the Human Mortality Database, in developed countries life expectancy at age 85 increased by only about one year between 1900 and 1960, but then increased by almost two years between 1960 and 1999. I call it the "100 curve": for citizens of developed countries the chance to live to 100 is now about 100 times higher than it was 100 years ago. In fact, if one projects the current trends according to the Human Mortality Database, most babies born since 2000 in developed countries will live to be 100.

James Vaupel, the founding director of the Max Planck Institute for Demographic Research, showed that the rate of increase in life expectancy is about 2.5 years per 10 years ("Demography", 2002). It means that every day our race's life expectancy increases by six hours. And Vaupel argues that life expectancy is likely to keep increasing.

These studies, however, often neglect facts of ordinary life. Since 1960 the conditions in which people live (especially urban people) have improved dramatically. For centuries people used to live with (and die because of) poor sanitation. The water supply of cities was chronically contaminated with sewage, garbage and carrion. Typhoid, dysentery and diarrhea were common. Outbreaks of smallpox, measles, polio, cholera, yellow fever, assorted plagues and even the flue killed millions before the invention of vaccines and the mandatory immunization programs of the last century. Before the 1960s polio would paralyze or kill over half a million people worldwide every year. Smallpox used to kill hundreds of thousands of Europeans annually (it was eradicated in 1979) and killed millions in the Americas after colonization. The World Health Organization estimates that measles has killed about 200 million people worldwide over the last 150 years (but almost nobody in developed countries in recent decades). Cholera killed 200,000 people in the Philippines in 1902-04, 110,000 in the Ukraine in 1910 and millions in India in the century before World War I. The flu killed at least 25 million people worldwide in 1918, four million in 1957 and 750,000 in 1968. These causes of death virtually disappeared from the statistics of developed countries in the last half century. After 1960 diseases are generally replaced by earthquakes, floods and hurricanes (and man-made famines in communist countries) in the list of the mass killers. The big exceptions, namely tuberculosis (more than one million deaths a year), AIDS (almost two million deaths a year) and malaria (more than 700,000 deaths a year), are now mostly confined to the under-developed countries that are not included in the studies on life expectancy (the World Bank estimates that 99% of deaths due to these three diseases occur in underdeveloped countries).

Another major factor that contributed to extending life expectancy is affordable professional health care. Health care used to be the responsibility of the family before it shifted towards the state. The state can provide more scientific health care, but it is expensive. Professional health care became affordable after World War II thanks to universal health care programs: France (1945), Britain (1948), Sweden (1955), Japan (1961), Canada (1972), Australia (1974), Italy (1978), Spain (1986), South Korea (1989), etc. Among major developed countries Germany (1889) is the only one that offered universal health care before World War II (and the USA is the only one that still does not have one in place).

After improved health care and reduced infectious disease rates, the economist Dora Costa's "Causes of Improving Health and Longevity at Older Ages" (2005) lists "reduced occupational stress" and "improved nutritional intake" as the other major factors that determine longevity. However, work stress is increasing for women, as they ascend the corporate ladder, and data on diets (e.g., obesity) seem to point in the opposite direction: people quit smoking, but now eat junk, and too much of it (and insert here your favorite rant against pesticides, hormone-raised meat and industrialized food in general).

Violent deaths have also decreased dramatically throughout the developed world: fewer and less bloody wars, and less violent crime. The rate of homicide deaths per 100,000 citizens is widely discussed in Steven Pinker's "The Better Angels of Our Nature" (2011). (Even in the USA where guns are widely available, and therefore violent crime kills exponentially more people than in Europe or Asia, the gun homicide rate decreased 49% between 1993 and 2013).

These factors certainly helped extend life expectancy in the developed world, but there is little improvement that they can still contribute going forward. In some cases one can even fear a regression. For example, no new classes of antibiotics have been introduced since 1987 whereas new pathogens are emerging every year, and existing bugs are developing resistance to current antibiotics. On the same day of March 2013 that a symposium in Australia predicted drugs to slow down the ageing process within a decade so that people can live to 150 years, the Chief Medical Officer for England, Dame Sally Davies, raised the alarm that antibiotics resistance may become a major killer in the near future. The Lancet, the British medical journal, estimated that in 2013 more than 58,000 babies died in India because they were born with bacterial infections that are resistant to known antibodies.

Drug-resistant tuberculosis killed an estimated 170,000 people in 2012. A 2016 report by the British government and the Wellcome Trust estimated that 700,000 people die every year from infections caused by drug-resistant pathogens. Instead of machine super-intelligence we should worry about biological super-bacteria.

The American Cancer Society calculated 1.6 million new cases of cancer and nearly 600,000 deaths in the USA in 2012, which means that the number of cancer deaths in the USA has increased by 74% since 1970. The World Health Organization's "World Cancer Report 2014" estimated that cancer cases will increase by 70 percent over the next 20 years.

The future promises more biomedical progress, and particularly therapies that may repair and reverse the causes of aging. This leads many to believe that human life can and will be extended dramatically, and maybe indefinitely.

However, health care has become too expensive for governments to continue footing the bill for the general population. Virtually every society in the developed world has been moving towards a larger base of elderly people and a smaller base of younger people who are supposed to pay for their health care. This equation is simply not sustainable. The professional health care that the average citizen receives may already have started to decline, and may continue to decline for a long time. It is just too expensive to keep the sick elderly alive forever for all the healthy youth who have to chip in. To compound the problem, statistics indicate that the number of people on disability programs is skyrocketing (14 million people in the USA in 2013, almost double the number of 15 years earlier). At the same time the tradition of domestic health care has largely been lost. You are on your own. This parallel development (unaffordable professional health care combined with the disappearance of domestic health care) is likely to reverse the longevity trend and lead to a worse (not better) chance of living a long life.

There have already been several times in the history of the Earth when intelligent life almost went extinct; and natural events such as plagues, volcano eruptions and meteorite crashes still constitute a threat to the fragile bodies that host our minds; not to mention the possibility of self-destruction by a nuclear war or by some kind of collective religious martyrdom in the name of some medieval superstition; i.e. by the stupidity of our so-called "intelligent" minds.

Furthermore, the rate of suicide has been increasing steadily in most developed societies, and, for whatever reason, it usually goes hand in hand with a decline in birth rates. Hence this might be an accelerating loop. The country with the oldest people is Japan. That is also one of the countries with the highest suicide rates of all, and most of the suicides are committed by elderly people. Getting very old does not make you very happy. In 2013 the Center for Disease Control (CDC) found that the suicide rate among middle-aged people in the USA had increased 28% in a decade (40% for white people) and that since 2009 suicide had become the 10th leading cause of death in the country, overtaking car accidents.

As all countries reach the point of shrinking health care and accelerating suicide rates, life expectancy will start to decline for the first time in centuries.

According to the National Center for Health Statistics of the US government, in 2015 the death rate rose for the first time in a decade. The CDC speculated that more people were dying from drug overdoses, suicide and Alzheimer's disease, but the death rate from heart disease, long in decline, was also slightly higher.

Jeanne Louise Calment died at the age of 122 in 1997. Since then no person in the developed world (where you can verify the age) has died at an older age. Even if you believed the claims from various supercentenarians in developing countries (countries in which no document can prove the age of very old people), you could hardly credit their achievement on technological or medical progress since those supercentennarians lived all their lives with virtually no help from technology or medicine. In other words, the real numbers tell us that in almost 20 years nobody has reached the age that someone reached in 1997 with the possible exception of people who lived in underdeveloped countries. It takes a lot of imagination to infer from this fact that we are witnessing a trend towards longer life-spans.

There is also a shift in value perception at work. The idea that the only measure of a life is the number of years it lasted, that dying of old age is "better" than, say, dying in a car accident at a young age, is very much grounded in an old society driven by the survival instinct: survive at all costs for as long as possible. As the (unconscious) survival instinct is progressively replaced by (conscious) philosophical meditation in modern societies, more and more people will decide that dying at 86 is not necessarily better than dying at 85. In the near future people may care more about other factors than the sheer number of years they lived. The attachment to life and the desire to live as long as possible is largely instinctive and irrational. As generations become more and more rational about life, longevity may not sound so attractive if one has to die anyway and be dead forever and be forgotten for the rest of eternity.

Then there are also many new habits that may contribute to creating a sicker species that will be more likely (not less likely) to die of diseases.

Most children in the developed world are now born to mothers (and fathers) aged 30 and older. As more parents delay childbearing, and the biological clock remains the same, the new generations are a veritable experiment (and sometimes literally a laboratory experiment). Fertility rates begin to decline gradually at age 30 and then decline exponentially, and to me that is nature's way of telling us when children should be born. In fact, babies born to older parents (as it is becoming common) have a higher risk of chromosome problems, as shown, for example, in a study led by Andrew Wyrobek and Brenda Eskenazi, "Advancing Age has Differential Effects on DNA Damage, Chromatin Integrity, Gene Mutations, and Chromosome Abnormalities in Sperm" (2006), and by a study led by Bronte Stone at the Reproductive Technology Laboratories, "Age thresholds for changes in semen parameters in men" (2013). Autism rates have risen 600 percent in the past two decades. While the age of the parents may not be the only cause, it is probably one significant cause. In Janie Shelton's and Irva Hertz-Picciotto's study "Independent and Dependent Contributions of Advanced Maternal and Paternal Ages to Autism Risk" (2010) the odds that a child would later be diagnosed with autism was 50% greater for a 40-year-old woman than for a woman between 25 and 29. To be fair, a study led by Mikko Myrskyla at the Max Planck

Institute for Demographic Research in Germany, "Maternal Age and Offspring Adult Health" (2012), reassured many older mothers that education is the main factor determining the future health of their babies.

Nobody knows its causes and it is difficult to speculate on the effects, but several studies from European nations seem to show that the quality of sperm has deteriorated during the last half of the previous century. I doubt that this bodes well for the physical and mental health of our offspring. For example, a study led by the epidemiologist Joelle LeMoal ("Decline in Semen Concentration and Morphology in a Sample of 26609 Men Close to General Population Between 1989 and 2005 in France", 2012) found that sperm concentration of young men decreased by nearly 30% in 17 years. If similar numbers showed up for environmental problems in a certain territory, we would immediately evacuate the place.

Last but not least, antibiotics, filtered water, cesarean-section childbirths and other environmental and behavioral aspects of modern life in developed countries greatly weaken the beneficial bacteria that constitute the physical majority of the cells of the human body. Vaccinations have been useful to prevent children from dying of horrible diseases, but now they are becoming mandatory for every possible disease (and for the mere possibility of a disease), thus creating weaker and weaker immune systems. Therefore health care itself (with its emphasis on vaccinations and antibiotics) may end up engineering weaker immune systems, which are much more likely to be defeated by unknown diseases than the unprotected immune system of our grandparents.

Personal mobility has greatly increased the chances that a deadly epidemic spreads worldwide killing millions of people.

Dana King's study "The Status of Baby Boomers' Health in the United States" (2013) seems to show that the "baby boomer" generation is less healthy than the previous generation. For example, twice as many baby boomers use a cane as did people of the same age in the previous generation. I personally have the feeling that the young people around me are less healthy than my generation was. Too many young people seem to have all sorts of physical problems and seem to get sick with every passing germ. They get food poisoning the moment they cross a border, and they start taking all sorts of pills in their 30s. I don't see how this bodes well for our race's longevity.

We now routinely eat genetically-manufactured food whose effects over the long term are yet to be determined.

One fears that most of the gains in life expectancy may have already occurred, and now the challenge will be to preserve them. I can't shake off the feeling that we are building a weaker and weaker species while creating a more and more dangerous world.

I was fascinated in 2016 when i saw at the same time two recent statistics, one (by the Pew Research Center) about the exponential growth of social media and the other (by the Center for Disease Control) about suicide rates. The number of suicides in the USA has been rising since 1999 in every age group and for both sexes. The rate of suicide has increased from 10.5 per 100,000 in 1999 to 13 per 100,000 in 2014. A study by David Lester, one of the world's experts in suicide statistics, published in 2002 in the Journal of the American Medical Association, noted a decline in the overall suicide rate between 1987 and 1997. So the suicide rate in the USA declined until 1997, then it started rising again. The coincidence is interesting: 1997 is the year that the first social network, Six Degrees, was launched.

Appendix: The Medium is the Brain

The World-wide Web has popularized the paradigm of navigating linked documents. This is an example of a process that introduces many distractions and inevitably reduces the depth of understanding (or, at least, increases the effort one has to make in order to stay focused). Generally speaking, the life of the individual who is permanently plugged into the network (hyperlink navigation, instant messages, live news) has a cost: the continuous shift of context and therefore of focus takes a cognitive toll on the brain.

Every time the brain has to reorient itself there is a cost in accessing long-time memory and organizing one's "thoughts". That brain gets trained for short attention spans. It is physically a different kind of brain from the brain that meditates and contemplates; from the brain that is capable of "deep understanding". The latter is trained by linear "texts" (or lectures or videos or whatever) that require the brain to remain on the same subject for a longer period of time and ransack long-term memory for all the appropriate resources to "understand" as much as possible. Brains that are trained to process linear texts comprehend more, remember more and, in my opinion, learn more, something already found in Erping Zhu's study "Hypermedia Interface Design" (1999). People who read linear text comprehend more, remember more, and learn more. Brains that are trained to switch focus all the time comprehend less, remember less and, possibly, learn less, as argued by Nicholas Carr in "The Shallows" (2010). This is due to the fact that it is "expensive" for the brain to transfer information from working memory to long-term memory (the "cognitive load"). Cognitive "overload" makes it difficult for the brain to decode and store information, and to create the appropriate links to pre-existing memories.

Guinevere Eden discussed how literacy reorganizes the brain at the physical level in "Development of Neural Mechanisms For Reading" (2003): reading and writing hijack a brain (so do other symbolic activities and art). Patricia Greenfield's study "Technology and Informal Education" (2009) shows that every medium develops some cognitive skills at the expense of others. Gary Small's "Patterns of Cerebral Activation during Internet Searching" (2009) proves how digital technology is rapidly and profoundly altering our brains. Betsy Sparrow's "Google Effects on Memory" (2011) shows how search engines change the way people use memory.

The medium that we use defines how the brain works. Ultimately, the medium physically changes our brain. The medium shapes the brain.

Every medium fosters some cognitive skills in the brain, but at the expense of others. There is a sort of zero sum of cognitive skills. A blind person improves smell and hearing. A videogame addict improves her visual-spatial skills but at the expense of other skills. The "focused" brain has skills that have been created by, for example, books, whereas the "switching" brain has skills that have been created by, for example, the Web.

The "switching" brain will lead to a more superficial society, in which brains are less and less capable of deep understanding. This is actually a process that has been going on for some centuries (if not millennia). At the time of Homer many people could memorize a lengthy poem. Before the invention of writing, brains had to memorize many more items than after the invention of writing. Before the invention of the specialist, people had to be experts in many fields of life, from carpentry to plumbing. After the invention of the society of specialists, we don't quite know how things work: we just know that by touching a switch or a lever something happens (a light comes on, a garage opens, a television set turns on, water comes out of a faucet). The history of civilization is a history of reducing the amount of cognitive skills required to survive. Civilizations have constantly been refining the process of finding and using knowledge at the expense of the process of storing and understanding knowledge. The Web-based society is simply a further step in this process, where navigating and multi-tasking prevail over deep understanding. We don't need to understand how things happen but just how to make things happen (e.g., if you want light, press a switch). Eventually, human brains may not be able to understand anything of the world that they "navigate" but will be able to do a lot more a lot faster in it.

This society of superficial brains will inevitably change the meaning of what is important. Science, literature and art were at the top of the hierarchy when deep understanding was important. Culture is not democratic at all. Academia decides what is more important and what is less important. In a society of superficial brains that don't need to understand much, it is debatable whether a classic poem is still more important than a pulp novel. The elite-controlled hierarchy of knowledge becomes pointless in a world of superficial brains.

The switching brain works in fundamentally different ways and inevitably creates a fundamentally different society of brains. Literacy reorganizes the brain at the physical level: reading and writing hijack a brain; browsing and searching hijack a brain too. Here are some of the changes in the way the switching brain works.

The Web has so much information that one does not need intelligence anymore to solve a problem: most likely the solution can be found by navigating hyperlinked pages on the Web. The new way to solve a problem is not to concentrate on the nature of the problem, study the dynamics of the system and then logically infer what the solution could be. The new way is to search the Web for the solution posted by someone who knows it. At one point Artificial Intelligence was trying to build "expert systems" that would use knowledge and inference to find solutions. The Web makes the amount of knowledge virtually infinite and reduces the inference required by problem solving to just searching the knowledge for an adequate match. No mathematical logic needed. We are evolving towards a less and less intelligent way of solving problems, albeit possibly a more and more effective way. The cognitive skill that we are losing is logical inference.

The combination of Web search and smartphones is also removing the need to think and argue about the truth of a statement: you can just "google" it and find the answer in a few seconds. There is no need to have a lengthy and emotional argument with a friend about who came first, the French or the USA revolution: just "google" it. Before the advent of the smartphone, one had to use all the brain's inferential skills and all the knowledge learned over a lifetime to guess the correct answer and to convince the audience. And the whole effort could easily lead to a wrong answer to be accepted by everybody. But that was a cognitive skill: rhetoric.

By the same token, there is no need to use a brain's orientation skills to find a place: just use the navigation system of the car or the smartphone. This removes the need to think and argue about whether to turn right or left. Before the advent of navigation systems, one had to use all the brain's inferential skills and all the knowledge learned over a lifetime to guess which way to go. And the whole effort could easily lead to picking a wrong direction. But that was a cognitive skill: orientation.

As our brain becomes more "superficial" it is likely that we also become more superficial in dealing with other individuals and with our world at large (family, friends, community, nation, our own life). One cognitive skill that may get lost in the age of "social networking" is precisely: socializing.

One skill that the switching brain is acquiring in place of the "focusing" skills is the skill of "posting" information. Before the Web only an elite was capable of producing content for the masses. The Web has created a large population of "prosumers", who are both passive consumers of content found on the Web and active producers of content for the Web (hats off to Alvin Toffler who coined the term in his 1980 book "The Third Wave", when the Internet was still an experiment). Social networking software, in particular, encourages people to post news about themselves, thus creating a diary read by (potentially) millions of people. This is fostering a cognitive skill about "marketing" yourself to the world, about how to present your personality and your life to the others.

The simple act of browsing the Web constitutes a new cognitive skill. The browser is becoming de facto a new organ of the body, an organ used to explore the virtual universe of the Web, just like a hand or an eye is used to explore the physical universe. This organ is generating a new sense just like the hand created the sense of touch and the eye created the sense of sight. This new sense implies a new function in the brain just like any sense implies a corresponding function in the brain.

The switching brain must also be refining another skill that has been evolving over the last century: choice. Before the invention of cable television and the multiplication of channels the viewer had little choice on what to watch. For example, there were only a few evening news programs (in some countries only one on the national channel). The whole country was watching the same news at the same time. There was no need for searching and choosing the news. Cable television and now the Web have multiplied the possible sources of news and made them available around the clock. The "superficial" brain may not want to delve deeply into any particular event but probably needs to be much more skilled at searching and choosing the news. Choice is also involved in social networking systems to decide what is worth discussing, what is worth knowing and what is worth telling others about yourself.

On the other hand, it is not only that tools influence our being, but also that our being influences tools. The story is as much about how tools use our brains as about how our minds use tools. Often people end up using a tool in a way that is not the one it was designed for. This is particularly obvious in the case of software applications, but also in the case of many technologies that became runaway successes "despite" the original intention of the inventors. So much so that different people may use the same tool in different manners for different purposes (e.g., Facebook). We express ourselves with the tools we have made as much as we see ourselves in the tools we have made.

The Web is the latest in a long series of new media that have shaped the human brain, starting with facial expression, language and writing. At each point some old skills were lost and some new skills were acquired. Your brain "is" the medium that shaped it. For better and for worse, you "are" the gadgets that you use.

Appendix: The Era of Objects

As we speculate on what will be the next stage of life's evolution, we underestimate the real innovation that has happened since the invention of intelligent life: objects. Life started building objects.

Life has not evolved much in the last ten thousand years, but objects have: there has been an explosive proliferation of objects.

We tend to focus on objects that mimic life (robots and the like) as candidates for replacing life as we know it, and in the long term that might well be correct, but the objects that have truly multiplied and evolved at an astonishing rate are the ordinary static objects that populate our homes, our streets and our workplaces. There are objects virtually for anything.

When we look at life's evolution, we tend to look at how sophisticated the brain has become, but we tend to underestimate what that "sophisticated" brain is built to do: make more objects. The chimp's brain and the human brain are not all that different, and the behavior of the two species (eating, sleeping, sex, and perhaps even consciousness) are not all that different from the viewpoint of a (non-chimp and non-human) external observer, but the difference in terms of objects that they make is colossal. The real evolution of the brain is in terms of the objects it can build.

What the human race has truly accomplished is to turn a vast portion of the planet into objects: paved streets and sidewalks, buildings, cars, trains, appliances, clothes, furniture, kitchenware, etc.

Our lives revolve around objects. We work to buy a car or a home, and our work mostly consists in building or selling objects. We live to use them (usually in conjunction with other objects), to place them somewhere (usually inside other objects), to clean them (using other objects), etc.

The fundamental property of life is that it dies. Everything that was alive is now dead, except for those that are dying now. For us the planet is just a vast cemetery. For objects, instead, this planet is a vast factory of objects because, before dying, each of us builds or buys thousands of objects that will survive us and that will motivate future generations to build and buy more objects.

It is neither living beings nor genes nor memes (ideas) that evolve and drive evolution on this planet: it is objects. Objects have evolved far faster than life or ideas. The explosive proliferation of objects is the one thing that would be visible to anyone playing the last ten thousand years of history on Earth. Everything else (politics, economics, natural disasters, etc) pales in comparison to the evolution and proliferation of objects. The human body has not changed much in 200,000 years. Ideas have changed but slowly. Objects, instead, have completely changed and keep changing rapidly.

For example, what caused the collapse of the Soviet Union (and of communism in general) was neither the Pope nor Afghanistan: it was consumerism. Soviet citizens wanted goods in their stores, lots of goods. They actually liked many features of the communist society (they still do) but they wanted the proliferation of goods that democratic/capitalist societies offer. It was all about objects. Hence the Soviet Union collapsed because it dared to challenge the domination of objects. For the same reason religions are declining everywhere: they are being replaced by philosophies of life that are more materialistic, i.e. that increase the evolution of objects.

Any system that challenges the absolute power of objects, or that doesn't contribute enough to the survival, proliferation and evolution of objects tends to lose. What benefits objects tends to succeed. Objects rule. Perhaps we are merely supposed to follow their orders, and that's the only meaning of life. We get annihilated if we dare contradict objects.

You may think that you are changing a car because you want a new car, but you can also see it the other way around: it is cars that want you to spend money that will go into making more and better cars.

In a sense, the consumer society is one stage in the evolution of objects, invented by objects in order to speed up their own evolution. Consumers are just the vehicle for objects to carry out their strategy of proliferation and domination.

Eventually objects will evolve into space stations and extraterrestrial colonies in order to expand outside this planet and begin the colonization of other parts of the universe in their quest to dominate all the matter that exists, until all matter in the universe will have been turned into objects by objects-creating beings like us (in clear defiance of the second law of Thermodynamics).

We are even turning our food into objects as we increasingly eat packaged food. The food system has changed more over the last 40 years than in the previous 40,000 years.

Shoes, refrigerators, watches and underwear are the real protagonists of history. Everything else is just a footnote to their odyssey.

(By the same token, i think that videogames play people, not the other way around).

Appendix: I Ran Out of Space

You can read more essays that didn't fit in this book at www.scaruffi.com/essays.html

A Timeline of Neuroscience

1590: Rudolph Goeckel's "Psychologia" introduces the word "psychology" for the discipline that studies the soul

1649: Pierre Gassendi's "Syntagma philosophiae Epicuri" argues that beasts have a cognitive life of their own, just inferior to humans

1664: Rene Descartes' "Treatise of Man" argues that the pineal gland is the main seat of consciousness (Great Minds Series):

1664: Thomas Willis' "Cerebral Anatomy" (1664) describes the different structures in the brain and coins the word "neurology"

1741: Emanuel Swedenborg's "The Economy of the Animate Kingdom" discusses cortical localization in the brain

1771: Luigi Galvani discovers that nerve cells are conductors of electricity

1796: Franz-Joseph Gall begins lecturing on phrenology, holding that mental faculties are localized in specific brain regions (of which 19 are shared with animals and 8 are exclusive to humans)

1824: Pierre Flourens' "Phrenology Examined" discredits Gall

1825: Jean-Baptiste Bouillaud's "Clinical and Physiological Treatise upon Encephalitis" describes patients who suffered brain lesions and lost their speech ability

1836: Marc Dax's "Lesions of the Left Half of the Brain Coincident With the Forgetting of the Signs of Thought" notes that aphasic patients (incapable of speaking) have sustained damage to the left side of the brain

1861: Paul Broca's "Loss of Speech, Chronic Softening and Partial Destruction of the Anterior Left Lobe of the Brain" single-handedly resurrects the theory of cortical localization of function

1865: Paul Broca's "Localization of Speech in the Third Left Frontal Convolution" suggests that the location of speech must be in the left hemisphere

1868: John Hughlings Jackson's "Notes on the Physiology and Pathology of the Nervous System" reports how damage to the right hemisphere impairs spatial abilities

1870: Eduard Hitzig and Gustav Fritsch discover the location of the motor functions in the brain

1873: Jean-Martin Charcot's "Lectures on the Diseases of the Nervous System" describes the neural origins of multiple sclerosis

1873: Camillo Golgi's "On the Structure of the Brain Grey Matter" describes the body of the nerve cell with a single axon and several dendrites

1874: Karl Wernicke determines that sensory aphasia (a loss of linguistic skills) is related to damage to the left temporal lobe

1874: Charles-Edouard Brown-Sequard's "Dual Character of the Brain" argues that education does not adequately target the right hemisphere

1876: John Hughlings Jackson discovers that loss of spatial skills is related to damage to the right hemisphere

1876: David Ferrier's "The Functions of the Brain" provides a map of brain regions specialized in motor, sensory and association functions

1890: Wilhelm His coins the word "dendrite"

1891: Santiago Ramon y Cajal proves that the nerve cell (the neuron) is the elementary unit of processing in the brain, receiving inputs from other neurons via the dendrites and sending its output to other neurons via the axon

1891: Wilhelm von Waldeyer coins the term "neuron" while discussing Santiago Ramon y Cajal's theory

1896: Albrecht von Kolliker coins the word "axon"

1897: Charles Sherrington coins the word "synapse"

1901: Charles Sherrington maps the motor cortex of apes

1903: Alfred Binet's "intelligent quotient" (IQ) test

1905: Keith Lucas demonstrates that below a certain threshold of stimulation a nerve does not respond to a stimulus and, once the threshold is reached, the nerve continues to respond by the same fixed amount no matter how strong the stimulus is

1906: Charles Sherrington's "The Integrative Action of the Nervous System" argues that the cerebral cortex is the center of integration for cognitive life

1911: Edward Thorndike's connectionism (the mind is a network of connections and learning occurs when elements are connected)

1921: Otto Loewi demonstrated chemical transmission of nerve impulses, proving that nerves can excite muscles via chemical reactions (notably acetylcholine) and not just electricity

1924: Hans Berger records electrical waves from the human brain, the first electroencephalograms

1924: Konstantin Bykov, performing split-brain experiments on dogs, discovers that severing the corpus callosum disables communications between the two brain hemispheres

1924: Hans Berger records electrical waves from the human brain, the first electroencephalograms

1925: Edgar Adrian shows that the message from one neuron to another neuron is conveyed by changes in the frequency of the discharge, the first clue on how sensory information might be coded in the neural system

1928: Otfried Foerster stimulates the brain of patients during surgery with electric probes

1933: Henry Dale coins the terms "adrenergic" and "cholinergic" to describe the nerves releasing the two fundamental classes of neurotransmitters, the adrenaline-like one and acetylcholine

1935: Wilder Penfield explains how to stimulate the brain of epileptic patients with electrical probes ("Epilepsy and Surgical Therapy")

1936: Jean Piaget's "The Origins of Intelligence in Children"

1940: Willian Van Wagenen performs "split brain" surgery to control epileptic seizures

1949: Donald Hebb's cell assemblies (selective strengthening or inhibition of synapses causes the brain to organize itself into regions of self-reinforcing neurons - the strength of a connection depends on how often it is used)

1951: Roger Sperry's "chemoaffinity theory" of synapse formation explains how the nervous system organizes itself during embryonic development via a genetically-determined chemical matching program

1952: Paul Maclean discovers the "limbic system"

1953: John Eccles' "The Neurophysiological Basis of Mind" describes excitatory and inhibitory potentials, the two fundamental changes that occur in neurons

1953: Roger Sperry and Ronald Meyers study the "split brain" and discover that the two hemispheres are specialized in different tasks

1953: Eugene Aserinsky discovers "rapid eye movement" (REM) sleep that corresponds with periods of dreaming

1954: Rita Levi-Montalcini discover nerve-growth factors that help to develop the nervous system, thus proving Sperry's chemoaffinity theory

1957: Vernon Mountcastle discovers the modular organization of the brain (vertical columns)

1959: Michel Jouvet discovers that REM sleep originates in the pons

1962: David Kuhl invents SPECT (single photon emission computer tomography)

1962: David Hubel's and Torsten Wiesel's "Receptive Fields, Binocular Interactive and Functional Architecture in the Cat's Visual Cortex"

1964: John Young proposes a "selectionist" theory of the brain (learning is the result of the elimination of neural connections)

1964: Paul Maclean's triune brain: three layers, each layer corresponding to a different stage of evolution

1964: Lueder Deecke and Hans-Helmut Kornhuber discover an unconscious electrical phenomenon in the brain, the Bereitschaftspotential (readiness potential)

1964: Benjamin Libet discovers that the readiness potential precedes conscious awareness by about half a second

1968: Niels Jerne's selectionist model of the brain (mental life a continuous process of environmental selection of concepts in our brain - the environment selects our thoughts)

1972: Raymond Damadian builds the world's first Magnetic Resonance Imaging (MRI) machine

1972: Jonathan Winson discovers a correlation between the theta rhythm of dreaming and long-term memory

1972: Godfrey Hounsfield and Allan Cormack invent computed tomography scanning or CAT-scanning

1973: Edward Hoffman and Michael Phelps create the first PET (positron emission tomography) scans that allow scientists to map brain function

1973: JeanPierre Changeux's "selective stabilization" theory

1977: Allan Hobson's theory of dreaming

1978: Gerald Edelman's theory of neuronal group selection or "Neural Darwinism"

1985: Michael Gazzaniga's "interpreter" (a module located in the left brain interprets the actions of the other modules and provides explanations for our behavior)

1988: Bernard Baars' "global workspace"

1989: Wolf Singer and Christof Koch discover that at, any given moment, very large number of neurons oscillate in synchrony and one pattern is amplified into a dominant 40 Hz oscillation (gamma synchronization)

1990: Seiji Ogawa's "functional MRI" measures brain activity based on blood flow

1994: Vilayanur Ramachandran proves the plasticity of the adult human brain

1996: Giacomo Rizzolatti discovers that the brain uses "mirror" neurons to represent what others are doing

1996: Rodolfo Llinas: Neurons are always active endlessly producing a repertory of possible actions, and the circumstances "select" which specific action is enacted

1997: Japan opens the Brain Science Institute near Tokyo

1998: Stanislas Dehaene's "global neuronal workspace"

2004: Zhuo-Hua Pan injects channelrhodopsin into the neurons of a live animal causing electrical activity with light (birth of optogenetics)

2005: Pascal Fries shows that neuronal communication is implemented via neuronal synchronization

2008: Nicholas Schiff's "Central Thalamic Contributions to Arousal. Regulation and Neurological Disorders of Consciousness"

2008: Yoichi Miyawaki's team at the ATR Computational Neuroscience Laboratories in Japan extracts visual memories from a person's brain

2009: The USA launches the Human Connectome Project to map the human brain

2011: Susumu Tonegawa's team creates false fear memories in mice using optogenetics

2012: Mark Mayford stores a mouse's memory of a familiar place on a microchip

2013: The European Union launches the Human Brain Project to computer-simulate the human brain

2013: Kwanghun Chung and Karl Deisseroth develop a technique to render brains transparent

2013: Marcello Massimini shows that conscious states involve widespread communication between different specialized regions of the brain

2013: Madeline Lancaster and Juergen Knoblich create brain tissue ("cerebral organoids")

2014: Sergiu Pasca creates three-dimensional neural cultures ("human cortical spheroids") from stem cells

2014: Yoshiki Sasai creates neurons of the cerebellum from embryonic stem cells

A Timeline of Artificial Intelligence

1935: Alonzo Church proves the undecidability of first order logic
1936: Alan Turing's Universal Machine ("On computable numbers, with an application to the Entscheidungsproblem")
1936: Alonzo Church's Lambda calculus
1941: Konrad Zuse's programmable electronic computer
1943: "Behavior, Purpose and Teleology" co-written by mathematician Norbert Wiener, physiologist Arturo Rosenblueth and engineer Julian Bigelow
1943: Kenneth Craik's "The Nature of Explanation"
1943: Warren McCulloch's and Walter Pitts' binary neuron ("A Logical Calculus of the Ideas Immanent in Nervous Activity")
1945: John Von Neumann designs a computer that holds its own instructions, the "stored-program architecture"
1946: The ENIAC, the first Turing-complete computer
1946: The first Macy Conference on Cybernetics
1947: John Von Neumann's self-reproducing automata
1948: Alan Turing's "Intelligent Machinery"
1948: Norbert Wiener's "Cybernetics"
1949: Leon Dostert founds Georgetown University's Institute of Languages and Linguistics
1949: William Grey-Walter's Elmer and Elsie robots
1950: Alan Turing's "Computing Machinery and Intelligence" (the "Turing Test")
1950: Claude Shannon's tree search
1951: Claude Shannon's maze-solving robots ("electronic rats")
1951: Karl Lashley's "The problem of serial order in behavior"
1952: First International Conference on Machine Translation organized by Yehoshua Bar-Hillel
1952: Ross Ashby's "Design for a Brain"
1954: Marvin Minsky's thesis on reinforcement learning
1954: Demonstration of a machine-translation system by Leon Dostert's team at Georgetown University and Cuthbert Hurd's team at IBM, possibly the first non-numerical application of a digital computer
1956: Allen Newell and Herbert Simon demonstrate the "Logic Theorist"
1956: Dartmouth conference on Artificial Intelligence
1957: Frank Rosenblatt's Perceptron
1957: Newell & Simon's "General Problem Solver"
1957: Noam Chomsky's "Syntactic Structures" (transformational grammar)
1958: John McCarthy's LISP programming language
1958: Oliver Selfridge's Pandemonium
1959: Arthur Samuel's Checkers, the world's first self-learning program

1959: John McCarthy and Marvin Minsky found the Artificial Intelligence Lab at the MIT

1958: John McCarthy's "Programs with Common Sense" focuses on knowledge representation

1959: Noam Chomsky's review of a book by Skinner ends the domination of behaviorism and resurrects cognitivism

1958: Yehoshua Bar-Hillel's "proof" that machine translation is impossible

1959: The industrial robot Unimate is deployed at General Motors

1960: Bernard Widrow's and Ted Hoff's Adaline ((Adaptive Linear Neuron or later Adaptive Linear Element) that uses the Delta Rule for neural networks

1960: Hilary Putnam's Computational Functionalism

1961: Melvin Maron's "Automatic Indexing"

1963 Irving John Good (Isidore Jacob Gudak) speculates about "ultraintelligent machines" (the "singularity")

1963 John McCarthy moves to Stanford and founds the Stanford Artificial Intelligence Laboratory (SAIL)

1964: IBM's "Shoebox" for speech recognition

1965: Ed Feigenbaum's Dendral expert system

1965: Lotfi Zadeh's Fuzzy Logic

1966: Leonard Baum's Hidden Markov Model

1966: Joe Weizenbaum's Eliza

1966: Ross Quillian's semantic networks

1967: Charles Fillmore's Case Frame Grammar

1968: Glenn Shafer's and Stuart Dempster's "Theory of Evidence"

1968: Peter Toma founds Systran to commercialize machine-translation systems

1969: First International Joint Conference on Artificial Intelligence (IJCAI) at Stanford

1969: Marvin Minsky & Samuel Papert's "Perceptrons" kill neural networks

1969: Roger Schank's Conceptual Dependency Theory for natural language processing

1969: Cordell Green's automatic synthesis of programs

1969: Stanford Research Institute's Shakey the Robot

1970: Albert Uttley's Informon for adaptive pattern recognition

1970: William Woods' Augmented Transition Network (ATN) for natural language processing

1971: Richard Fikes' and Nils Nilsson's STRIPS planner

1971: Ingo Rechenberg's "Evolution Strategies"

1972: Alain Colmerauer's PROLOG programming language

1972: Bruce Buchanan's MYCIN

1972: Hubert Dreyfus's "What Computers Can't Do"

1972: Terry Winograd's Shrdlu

1973: "Artificial Intelligence: A General Survey" by James Lighthill criticizes Artificial Intelligence for over-promising

1973: Jim Baker applies the Hidden Markov Model to speech recognition

1974: Marvin Minsky's Frame

1974: Paul Werbos' Backpropagation algorithm for neural networks

1975: John Holland's genetic algorithms
1975: Roger Schank's Script
1975: Raj Reddy's team at Carnegie Mellon University develops three speech-recognition systems (Bruce Lowerre's "Harpy", Hearsay-II and Jim Baker's Dragon)
1976: Fred Jelinek's "Continuous Speech Recognition by Statistical Methods"
1976: Doug Lenat's AM
1976: Richard Laing's paradigm of self-replication by self-inspection
1979: David Marr's theory of vision
1979: Drew McDermott's non-monotonic logic
1979: William Clancey's Guidon
1980: Intellicorp, the first major start-up for Artificial Intelligence
1980: John McDermott's Xcon
1980: John Searle' "Minds, Brains, and Programs" on the "Chinese Room"
1980: Kunihiko Fukushima's Convolutional Neural Networks
1980: McCarthy's Circumscription
1981: Danny Hillis' Connection Machine
1981: Hans Kamp's Discourse Representation Theory
1982: Japan's Fifth Generation Computer Systems project
1982: John Hopfield describes a new generation of neural networks, based on a simulation of annealing
1982: Judea Pearl's "Bayesian networks"
1982: Teuvo Kohonen's Self-Organized Maps (SOM) for unsupervised learning
1982: The Canadian Institute for Advanced Research (CIFAR) establishes Artificial Intelligence and Robotics as its very first program
1983: Geoffrey Hinton's and Terry Sejnowski's Boltzmann machine for unsupervised learning
1983: Gerard Salton's "Introduction to Modern Information Retrieval"
1983: John Laird and Paul Rosenbloom's SOAR
1984: Valentino Braitenberg's "Vehicles"
1986: David Rumelhart's "Parallel Distributed Processing" rediscovers Werbos' backpropagation algorithm
1986: Paul Smolensky's Restricted Boltzmann machine
1986: Barbara Grosz's "Attention, Intentions, and the Structure of Discourse"
1987: Chris Langton coins the term "Artificial Life"
1987: Hinton moves to the Canadian Institute for Advanced Research (CIFAR)
1987: Marvin Minsky's "Society of Mind"
1987: Rodney Brooks' robots
1987: Stephen Grossberg's Adaptive Resonance Theory (ART) for unsupervised learning
1988: Toshio Fukuda's self-reconfiguring robot CEBOT
1988: Fred Jelinek's team at IBM publishes "A statistical approach to language translation"
1988: Hilary Putnam: "Has artificial intelligence taught us anything of importance about the mind?"

1988: Philip Agre builds the first "Heideggerian AI", Pengi, a system that plays the arcade videogame Pengo

1988: Fred Jelinek's team at IBM publishes "A Statistical Approach to Language Translation"

1989: Yann LeCun's "Backpropagation Applied to Handwritten Zip Code Recognition"

1989: Chris Watkins' Q-learning

1990: Robert Jacobs' "mixture-of-experts" architecture

1990: Carver Mead describes a neuromorphic processor

1990: Peter Brown at IBM implements a statistical machine translation system

1990: Ray Kurzweil's book "Age of Intelligent Machines"

1992: Thomas Ray develops "Tierra", a virtual world

1992: Hava Siegelmann's and Eduardo Israel's Recurrent Neural Networks (RNNs)

1994: The first "Toward a Science of Consciousness" conference in Tucson, Arizona

1995: Geoffrey Hinton's Helmholtz machine

1995: Vladimir Vapnik's "Support-Vector Networks"

1996: David Field & Bruno Olshausen's sparse coding

1997: Sepp Hochreiter's and Jeurgen Schmidhuber's LSTM model

1997: IBM's "Deep Blue" chess machine beats world champion, Garry Kasparov

1998: Two Stanford students, Larry Page and Russian-born Sergey Brin, launch the search engine Google

1998: Thorsten Joachims' "Text Categorization With Support Vector Machines"

1998: Yann LeCun's second generation Convolutional Neural Networks

2000: Cynthia Breazeal's emotional robot, "Kismet"

2000: Seth Lloyd's "Ultimate physical limits to computation"

2001: Juyang Weng's "Autonomous mental development by robots and animals"

2001: Nikolaus Hansen introduces the evolution strategy called "Covariance Matrix Adaptation" (CMA) for numerical optimization of non-linear problems

2002: iRobot's Roomba

2003: Hiroshi Ishiguro's Actroid, a robot that looks like a young woman

2003: Jackrit Suthakorn and Gregory Chirikjian at Johns Hopkins University build an autonomous self-replicating robot

2003: Yoshua Bengio's "Neural Probabilistic Language Model"

2003: Tai-Sing Lee's "Hierarchical Bayesian inference in the visual cortex"

2004: Ipke Wachsmuth's conversational agent "Max"

2004: Mark Tilden's biomorphic robot Robosapien

2005: Andrew Ng at Stanford launches the STAIR project (Stanford Artificial Intelligence Robot)

2005: Boston Dynamics' quadruped robot "BigDog"

2005: Hod Lipson's "self-assembling machine" at Cornell University

2005: Honda's humanoid robot "Asimo"

2005: Pietro Perona's and Fei-Fei's "A Bayesian Hierarchical Model for Learning Natural Scene Categories"

2005: Sebastian Thrun's driverless car Stanley wins DARPA's Grand Challenge

2006: Geoffrey Hinton's Deep Belief Networks (a fast learning algorithm for restricted Boltzmann machines)

2006: Osamu Hasegawa's Self-Organising Incremental Neural Network (SOINN), a self-replicating neural network for unsupervised learning

2006: Robot startup Willow Garage is founded

2007: Yoshua Bengio's Stacked Auto-Encoders

2007: Stanford unveils the Robot Operating System (ROS)

2008: Adrian Bowyer's 3D Printer builds a copy of itself

2008: Cynthia Breazeal's team at the MIT's Media Lab unveils Nexi, the first mobile-dexterous-social (MDS) robot

2008: Dharmendra Modha at IBM launches a project to build a neuromorphic processor

2009: FeiFei Li's ImageNet database of human-tagged images

2010: Daniela Rus' "Programmable Matter by Folding"

2010: Lola Canamero's Nao, a robot that can show its emotions

2010: Quoc Le's "Tiled Convolutional Networks"

2010: Andrew Ng's "Learning Continuous Phrase Representations and Syntactic Parsing with Recursive Neural Networks"

2010: The New York stock market is shut down after algorithmic trading has wiped out a trillion dollars within a few seconds.

2011: IBM's Watson debuts on a tv show

2011: Nick D'Aloisio releases the summarizing tool Trimit (later Summly) for smartphones

2011: Osamu Hasegawa's SOINN-based robot that learns functions it was not programmed to do

2012: Rodney Brooks' hand programmable robot "Baxter"

2012: Alex Krizhevsky and Ilya Sutskever demonstrate that deep learning outperforms traditional approaches to computer vision processing 200 billion images during training

2013: Volodymyr Mnih's Deep Q-Networks

2014: Vladimir Veselov's and Eugene Demchenko's program Eugene Goostman, which simulates a 13-year-old Ukrainian boy, passes the Turing test at the Royal Society in London

2014: Fei-Fei Li's computer vision algorithm that can describe photos

2014: Alex Graves, Greg Wayne and Ivo Danihelka publish a paper on "Neural Turing Machines"

2014: Jason Weston, Sumit Chopra and Antoine Bordes publish a paper on "Memory Networks"

2014: Microsoft demonstrates a real-time spoken language translation system

2015: Over 1,000 high-profile Artificial Intelligence scientists sign an open letter calling for a ban on "offensive autonomous weapons"

2016: Google's AlphaGo beats Go master Lee Se-dol

Readings on the Singularity

Barricelli, Nils: "Suggestions for the Starting of Numeric Evolution Processes Intended to Evolve..." (1987)

Brynjolfsson, Erik & McAfee, Andrew: "Race Against the Machine" (2012)

Bostrom, Nick: "Are we living in a simulation?" (2003)

Bostrom, Nick: "Superintelligence" (2014)

Carr, Nicholas: "Utopia Is Creepy and other Provocations" (2016)

Chalmers, David: "The Singularity - A Philosophical Analysis" (2010)

Cooke, Conrad: "Automata Old and New" (1893)

Domingos, Pedro: "The Master Algorithm" (2015)

Eden, Amnon & others: "The Singularity Hypotheses" (2013)

Geraci, Robert: "Apocalyptic A.I." (Oxford, 2010)

Good, Irving John: "Speculations Concerning the First Ultraintelligent Machine" (1964)

Goodfellow, Ian & Bengio, Yoshua: "Deep Learning" (2016)

Kaku, Michio: The Future of Intelligence (Doubleday, 2014)

Kurzweil, Ray: "The Age of Intelligent Machines" (1990)

Lanier, Jaron: "The Singularity Is Just a Religion for Digital Geeks" (2010)

Moravec, Hans: "Mind Children: The Future of Robot and Human Intelligence" (1988)

Muehlhauser, Luke: "When Will AI Be Created?" (2013)

Noble, David: "The Religion of Technology" (1997)

Scaruffi, Piero: "Thinking about Thought" (2015)

Vinge, Vernon: "The Singularity" (1993)

IEEE Spectrum's "Special Report: The Singularity" (2008)

The Journal of Consciousness Studies's Volume 19/7-8 (2012):

    Corabi & Schneider: "The metaphysics of uploading"

    Dainton, Barry: "On singularities and simulations"

    Hutter, Marcus: "Can intelligence explode?"

    Steinhart, Eric: "The singularity: Beyond philosophy of mind"

Alphabetical Index

Praise for "Intelligence is not Artificial":

"When it comes to any technology it is crucial to consider its reality, and not simply the hype put out by the media, or by the individuals and industries that are poised to profit by those techonologies. 'Intelligence is Not Artificial' does a great job cutting through the mountains of propaganda to present sobering and refreshing realities of its subject."
    (Mitch Altman, virtual-reality pioneer and founder of one of the earliest hackerspaces, Noisebridge)

"Creative thinking, outside the box. "Intelligence is not Artificial" is a well thought-out counterpoint to the contemporary stream of artificial intelligence prophecies and assertions."
    (Eric Gordon, Consulting Professor at Stanford University, Senior Advisor at Skyline Ventures, President of Palantir Consulting)

"I loved already the premise: like so many past ages, we inflate our importance in the overall arch of history. And i love this book's wet blanket reality thrown over the arrogant self importance of the AI scientist/hipster class."
    (John Law, cofounder of the Burning Man Festival)

www.ingramcontent.com/pod-product-compliance
Lightning Source LLC
Chambersburg PA
CBHW060426200326
41518CB00009B/1502